The Biotech Revolution:

Impact on Science Education in America

Ken R. Harewood

authorHOUSE

AuthorHouse™
1663 Liberty Drive
Bloomington, IN 47403
www.authorhouse.com
Phone: 833-262-8899

© 2020 Ken R. Harewood. All rights reserved.

No part of this book may be reproduced, stored in a retrieval system, or transmitted by any means without the written permission of the author.

Published by AuthorHouse 09/23/2020

ISBN: 978-1-7283-7099-6 (sc)
ISBN: 978-1-7283-7097-2 (hc)
ISBN: 978-1-7283-7098-9 (e)

Library of Congress Control Number: 2020916227

Print information available on the last page.

Any people depicted in stock imagery provided by Getty Images are models, and such images are being used for illustrative purposes only.
Certain stock imagery © Getty Images.

This book is printed on acid-free paper.

Because of the dynamic nature of the Internet, any web addresses or links contained in this book may have changed since publication and may no longer be valid. The views expressed in this work are solely those of the author and do not necessarily reflect the views of the publisher, and the publisher hereby disclaims any responsibility for them.

CONTENTS

Acknowledgement vii
Introduction ix

Chapter 1   DNA's 50th Anniversary 1
Chapter 2   The Central Dogma of Molecular Biology 11
Chapter 3   The Biotechnology Revolution 21
Chapter 4   The Human Genome Sequence 25
Chapter 5   America's Global Leadership in Science............. 31
Chapter 6   The American Education System 39
Chapter 7   Integrating American Higher Education 49
Chapter 8   Classifying Colleges and Universities 55
Chapter 9   Funding Colleges and Universities 63
Chapter 10   Science for the Citizen 73
Chapter 11   Teaching Versus Research 81
Chapter 12   Science in the New Millennium........................ 91

ACKNOWLEDGEMENT

To my grandchildren Brandon, Lauryn, Jaden, Mia, Danielle, and Evan. This book is intended to let you know how science influences your lives and the lives of everyone on our planet.

Graphic Design by my granddaughter Lauryn Fauntroy. Editing by my niece Gia Harewood, Ph.D.

INTRODUCTION

For more than forty years, biotechnology and information technology have been the principal drivers of economic and workforce development in America. In the twentieth century, unprecedented advances in basic and applied science gave birth to what has been called the knowledge economy, a development that placed new demands on the nation's colleges and universities. For centuries, this diverse system of institutions has served America well by educating countless generations of highly creative scientists.

In the twenty-first century, America's colleges and universities continue to sponsor research conducted by thousands of the nation's brightest and best scientists. These innovators, assisted by teams of graduate students and post-doctoral fellows, conduct research, publish their results in peer-reviewed journals, and present their findings at national and international meetings. Those activities speak eloquently to the contributions colleges and universities make to America's intellectual and economic development.

Funding for the research conducted by America's higher education institutions is provided through grant and contract mechanisms. The National Institutes of Health (NIH) is the largest public sector agency established by Congress to address the nation's biomedical and public health needs. It fulfills this mission by funding research projects deemed most likely to expand fundamental knowledge of biological systems, particularly those complex molecular events that enable scientists to gain a better understanding of the pathophysiology of human diseases.

A competitive grants and contracts process is the primary mechanism NIH uses to identify and fund proposals that are best aligned with the agency's goals and objectives.

The pharmaceutical industry (Big Pharma) is by far the leading private sector player involved in advancing America's research competitiveness. This group of highly diversified companies fund multidisciplinary research conducted intramurally by thousands of dedicated employees. Working under severe time constraints, pharmaceutical scientists engage in a relentless effort to develop impressive portfolios of commercially useful products and processes for their corporate employers.

When public and private sector inputs are tallied, America spends trillions of dollars annually supporting basic and applied research, with private sector contributions to new product/process development exceeding, by far, outputs from all government agencies combined.

This model for funding research and development (R&D) is quintessentially American. For decades, it has been the source of the nation's unrivaled leadership in aeronautics, manufacturing, and space exploration.

In the mid-twentieth century, after the structure of the genetic material was revealed by American scientist James Watson, funding for public and private sector research increased exponentially, resulting in the creation of an extraordinary number of companies dedicated to the production of novel biological products.

The explosion of new companies prompted Congress to act, and regulations addressing the ethical, legal, and social implications (ELSI) of manipulating the human genome were written into law.

What Congress failed to do then was craft legislation to help America's higher education institutions adjust to the demands of a post-genomic world, particularly after the draft sequence of the human genome was published at the beginning of this century.

This book chronicles events leading to Watson's report on the structure of deoxyribonucleic acid (DNA) in 1953. It draws attention to the fiftieth anniversary celebration of that revolutionary discovery, and illustrates how emphasis on applied science propelled America to a position of global leadership in science and technology since the dawn of the nineteenth century.

It examines the nation's education system, revealing how profoundly science influences the lives of everyday Americans. The narrative draws attention to legislation passed by Congress that created the land-grant colleges in the nineteenth century. It recounts the protracted struggle that ensued when the newly created land-grant colleges refused to educate the children of formerly enslaved persons—actions that prompted Congress to enact rules that extended land-grant status to a group of institutions referred to as Historically Black Colleges and Universities (HBCUs).

Elevating HBCUs to land-grant status resulted in the creation of a two-tiered higher education system—one devoted to educating Whites, and the other dedicated to serving Blacks.

This book also highlights some of the Draconian practices that state and federal agencies resorted to so that they could perpetuate the practices of racial segregation. And it recounts how Blacks used the courts to put an end to discrimination in the public education system.

Information is also provided that shows the extent to which private entities contributed to the development of America's higher education system. One such organization was the Carnegie Commission on Higher Education that created the Carnegie Classification System for Colleges and Universities. It was through periodic Carnegie Classification System reports that Americans really began to learn about the vast assortment of higher education institutions involved in helping to meet the nation's science and technology needs.

The other organizations highlighted in this book are the agencies that conduct periodic assessments of America's higher education institutions. These private accrediting agencies are responsible for ensuring that all colleges and universities remain in full compliance with standards authorized by Congress and enforced by the U.S. Department of Education (DoED).

Inclusion of the Carnegie Classification System and private accrediting agencies in this book is intended to show how reports provided by private organizations impact decision making by students, parents, funding agencies, personnel managers, and other higher education stakeholders.

One of the major issues this book addresses is the pace at which information technology is changing the landscape of science education in America's colleges and universities. The proliferation of distance education and online courses in science and technology is cited as one example of how, in the twenty-first century, some prestigious higher education institutions are providing students with convenient access to degree offerings at much lower cost. Because this trend is likely to continue, the science curriculum reform initiative recommended in the final chapter addresses the urgent need to merge the best elements of traditional didactic teaching with the most effective online strategies currently in use.

Finally, publication of this book will remind a proud and prosperous citizenry of the important roles science and technology play in improving the lives of all Americans. Policy makers should embrace the call for science curriculum reform as a national imperative. Crafting a curriculum that meets the education and training needs of the twenty-first century is a major challenge. Failing to do so could negatively impact the contributions that America's higher education institutions make to the advancement of science and technology for generations to come.

CHAPTER 1

DNA's 50th Anniversary

Almost seventeen years ago, some of the world's most accomplished scientists gathered at Cold Spring Harbor Laboratories (CSHL) on Long Island to celebrate the fiftieth anniversary of one of the greatest scientific advances of the twentieth century—the discovery of the three-dimensional structure of DNA. That symposium was the culmination of a yearlong series of events organized to pay tribute to Dr. James Watson—the scientist whose one-page manuscript published in the journal *Nature* revolutionized thinking about how genetic information is organized within living cells.

I was enrolled in middle school when Watson's manuscript was published in 1953, and not surprisingly, my science teacher, for whom I have enormous respect, spent little classroom time talking about its implications. Doing so would have placed him at odds with an education system that rigorously embraced didactic teaching over research. Accordingly, my lectures were textbook focused. They were designed to assess how well I could memorize the major laws of biology, chemistry and physics, as well as the names of scientists credited with establishing those laws. A discussion of Watson's publication in that setting would have been highly unusual. It was not until graduate school that I learned about his remarkable discovery, and how it changed the way science would be taught and practiced by future generations.

Fifty years after Watson's groundbreaking report, a new generation of scientists was in attendance at the CSHL meeting. Those familiar with Watson's past fully understood why CSHL was chosen as the venue for such an important event. They knew it was the place where Watson gave his first public lecture after publishing that seminal report on DNA. Some of the attendees were also aware of the fact that CSHL was the institution where Watson spent most of his adult life.

On that day in 2003, as I sat in my office at North Carolina Central University (NCCU), I couldn't help thinking about the life of James Watson the scientist. I didn't know him when I first read his manuscript or wrote a term paper on the structure of DNA for my undergraduate biochemistry class. Later in my career, I had the good fortune of meeting him. I found him to be the consummate storyteller. He could mesmerize an audience with his uncanny ability to simplify the most complex data, while keeping everyone on the edge of their seats with interesting anecdotal remarks about his colleagues and competitors. Whenever I talked to students about him, I told them that passion, persistence and pushiness were among his greatest strengths.

James Dewey Watson, the individual at the center of that fiftieth anniversary celebration in New York, is without doubt a genuine

American icon. His creativity and ingenuity are indisputable, and his contributions to science continue to be the subject of numerous articles and books. What is most remarkable about his story is the fact that in his early life he was not considered to be an exceptional student. The picture of him that sticks in my mind is one of a young man who during his childhood was primarily interested in birds. In his book entitled *The Double Helix* he provides a riveting personal account of the events and personalities associated with his quest to solve the structure of DNA.

Watson obtained his bachelor's degree from the University of Chicago where he majored in Zoology. As a graduate student in genetics at Indiana University, he came under the influence of Dr. Salvador Lauria, a famous bacteriologist who was largely responsible for Watson's obsession with the study of viruses. Like Lauria, he felt that those submicroscopic particles that invade plant, animal and bacterial cells were ideal candidates for experiments he was proposing to carry out for his doctoral thesis.

The particular agent that he elected to study was a virus referred to as a bacteriophage because of its propensity to infect bacterial cells. He believed that if he bombarded the bacteriophage with a physical agent like X-rays, he might be able to alter its DNA in a way that could provide clues about that molecule's role in regulating the agent's biological functions. His assumption seemed completely reasonable, because he knew that DNA comprised approximately fifty percent of the mass of the bacteriophage.

The rationale for Watson's approach was consistent with ideas advanced by Austrian physicist Erwin Schrodinger in his book entitled *What is Life*. The central theme of Schrodinger's book was that living cells contain basic informational units called genes, and that learning how these entities work would shed light on how genetic characteristics are passed from parent to offspring. Watson was deeply influenced by Schrodinger's ideas and he wanted to learn as much as he could about genes and DNA. He was also aware of a report by Avery, McCarty and MacLeod, three investigators working at the Rockefeller Institute in New York, that showed hereditary traits could be transmitted from one bacterial cell to another using purified DNA. This and other peer-reviewed publications from that era strengthened his conviction that DNA might be the hereditary material.

In 1950, after completing doctoral studies, Watson went to Copenhagen to begin postdoctoral training. His project focused on analyzing the chemical makeup of DNA. While in Denmark, he took a trip to Naples, Italy where he met Maurice Wilkins, a British physicist who was an expert at using X-rays to study DNA. After that brief encounter with Wilkins, Watson felt that his work on DNA could benefit significantly if he completed his research in England—a place where detailed studies on the structure of large molecules were already underway.

With Lauria's help, Watson secured a place at the Cavendish Laboratory in Cambridge where he joined a small group of physicists and chemists working on the three-dimensional structure of large proteins. In 1951, when he arrived in Cambridge he was twenty-three years old!

One of the investigators Watson met at the Cavendish Laboratory was Dr. Max Perutz, an Austrian scientist who was an expert at using X-rays to study the three-dimensional structure of proteins. It was in Perutz's laboratory where he began to interact with Dr. Francis Crick—a physicist who eventually became one of his principal collaborators. Crick's interest in studying large molecules, combined with his quick, analytical mind proved to be invaluable assets that Watson drew on during the time he spent at Cavendish.

With Crick's assistance, Watson was able to reconnect with Maurice Wilkins, the analytical scientist he met in Copenhagen. Wilkins worked at Kings College in London, and Watson would make the fifty miles trip from Cavendish to Kings College on numerous occasions. Through interacting with Wilkins and his research assistant Rosalind Franklin, Watson gained access to an impressive collection of X-ray diffraction images of DNA.

There are different accounts of the exchanges that took place between Watson, Crick, Wilkins, Franklin and their collaborators from 1951 to 1953, but what is indisputable is the fact that during that time, Watson studied Franklin's X-ray images, and together with Crick he pulled off a feat of extraordinary proportion—solving DNA's three dimensional structure.

In 1962, James Watson, Francis Crick and Maurice Wilkins were awarded the Nobel Prize in Medicine and Physiology for their incredible contribution to science!

I chose to re-visit the Watson story at the beginning of this book because it shows how important creativity and innovation are as drivers of the scientific discovery process. I also wanted to use that story to show how valuable institutional culture can be in fostering and promoting intellectual rigor. There is no doubt that

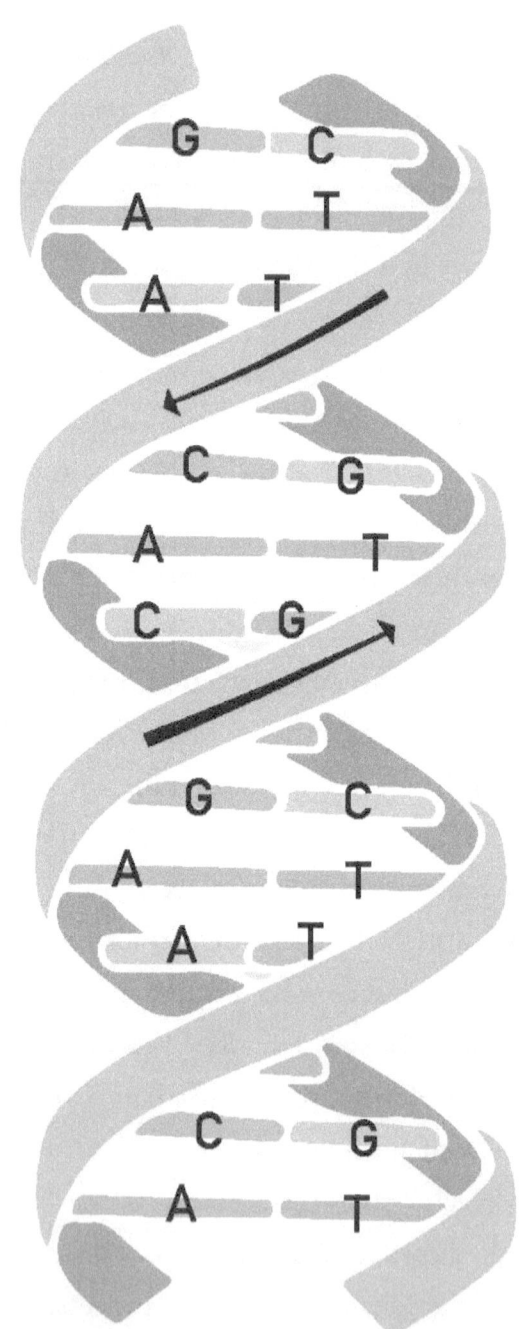

DNA Double Helix

it was the combination of Watson's ingenuity with the exacting standards that permeated the culture at Cavendish that led to one of the most revolutionary discoveries of the twentieth century.

Like Cavendish, CSHL—the site of DNA's fiftieth anniversary celebration in 2003—has advanced the careers of many young American scientists. My first visit there was in 1971—the year I joined Pfizer Inc. to work in the Cancer Research Department which, at that time, was located in the town of Maywood, New Jersey. Our program was housed in the John L. Smith Memorial for Cancer Research—a research institute established by the Pfizer Foundation and named after a former president of the company who died from leukemia. The primary objective of its operation was to study the etiology of human cancer, particularly leukemia.

I was thrilled to learn that my application to attend that year's CSHL symposium on viruses was approved by the conference organizers. The meeting was held on Mother's Day weekend, and it attracted a large gathering of highly accomplished investigators to that picturesque campus on Long Island. I remember every year the program would include at least one groundbreaking presentation. I did not realize then that my annual trek to CSHL would provide me with some of the most exciting and enriching experiences of my scientific career.

Generations of American and international scientists attended those CSHL annual symposia. Several enrolled in special courses advertised in slick brochures, while others published their results in the Laboratory's journals. A privileged group was offered the opportunity to participate in CSHL Banbury conferences. I found the latter to be particularly stimulating because of their small size and the unique opportunity they afforded for one-on-one interaction with some of the most accomplished international scientists.

During my many visits to CSHL, I took time to delve into the history of that prestigious institution. I learned that it was founded at the end of the nineteenth century by the Brooklyn Institute of Arts and Science. Its original mission was to train high school and college teachers in the discipline of marine biology. Later, that responsibility was expanded to include genetics. By 1945, as interest in genetics increased, CSHL began to offer a special course on viruses. That Phage Course as it was called, was taught by the celebrated Nobel Laureate Max Delbruck, and it helped to stimulate interest in the sub-discipline of biology known as bacteriology.

In 1968, after a succession of changes in the leadership at CSHL, James Watson was named director. He served there for decades, becoming chancellor in 2003 and chancellor emeritus in 2008. Under his guidance, the Laboratory sponsored annual symposia, published several peer-reviewed journals, and increased its roster of special courses.

Today, every time I think about the challenges that primary, secondary and tertiary institutions face as they grapple with the formidable task of educating the next generation of American scientists, I think of CSHL and of James Watson. I reflect on the nexus between creativity and institutional culture, and I acknowledge the role that both play in promoting scientific excellence. I am heartened by these thoughts and remain optimistic that science will always continue to be a major driver of American exceptionalism.

CHAPTER 2

The Central Dogma of Molecular Biology

Predictably, Watson and Crick's publication on the structure of DNA triggered unprecedented advances in basic science. Beginning in the 1960s and continuing over the next two decades, American scientists used information from Watson's report on the structure of DNA to gain a better understanding of how genetic information is organized, stored and transmitted within living cells.

Building on painstaking analytical studies carried out by Erwin Chargaff of Columbia University, four types of chemical building blocks called nucleotides were identified in each strand of the DNA molecule. These chemical units designated Adenine, Thymine, Cytosine, and Guanine are represented by the letters A, T, C, and G, respectively.

The two strands of each DNA molecule were shown to be oriented in an anti-parallel fashion with the nucleotide bases interacting in pairs, A pairing with T and G with C. One of the most fascinating things revealed about how the nucleotide bases are arranged along the linear stretch of each DNA chain is that they occur as triplets commonly referred to as codons. Each three-letter codon represents a 'word' in the genetic lexicon, and these 'words' make sense in much the same way that words formed from the twenty-six letters of the alphabet give meaning to the English language.

In 1961, armed with a growing body of data on the chemical, physical and biological properties of DNA, NIH scientist Dr. Marshall Nirenberg published his pioneering work showing how proteins are assembled inside living cells. It marked a significant step toward understanding the molecular triggers of cell growth and differentiation. Nirenberg's finding helped to 'crack' the genetic code, and for that, he and his colleagues Drs. Gobind Khorana and Robert Holley were awarded the Nobel Prize in Physiology and Medicine in 1968.

Collectively, these discoveries enabled scientists to construct a stepwise process revealing how information stored in DNA is faithfully replicated during cell division and utilized to carryout the basic functions of living organisms.

The process consists of three steps. During the first step, DNA enters its replication cycle, and as the molecule unwinds from one end, an enzyme whose properties are well understood duplicates each exposed strand.

The second step—transcription—is triggered by a different enzyme, one that has the capacity to bind to the unwound DNA strand, faithfully copying the genetic information encoded in its linear sequence. The product is a single stranded molecule called messenger ribonucleic acid (mRNA). A key feature of mRNA is its complimentarity to the DNA strand that served as its template.

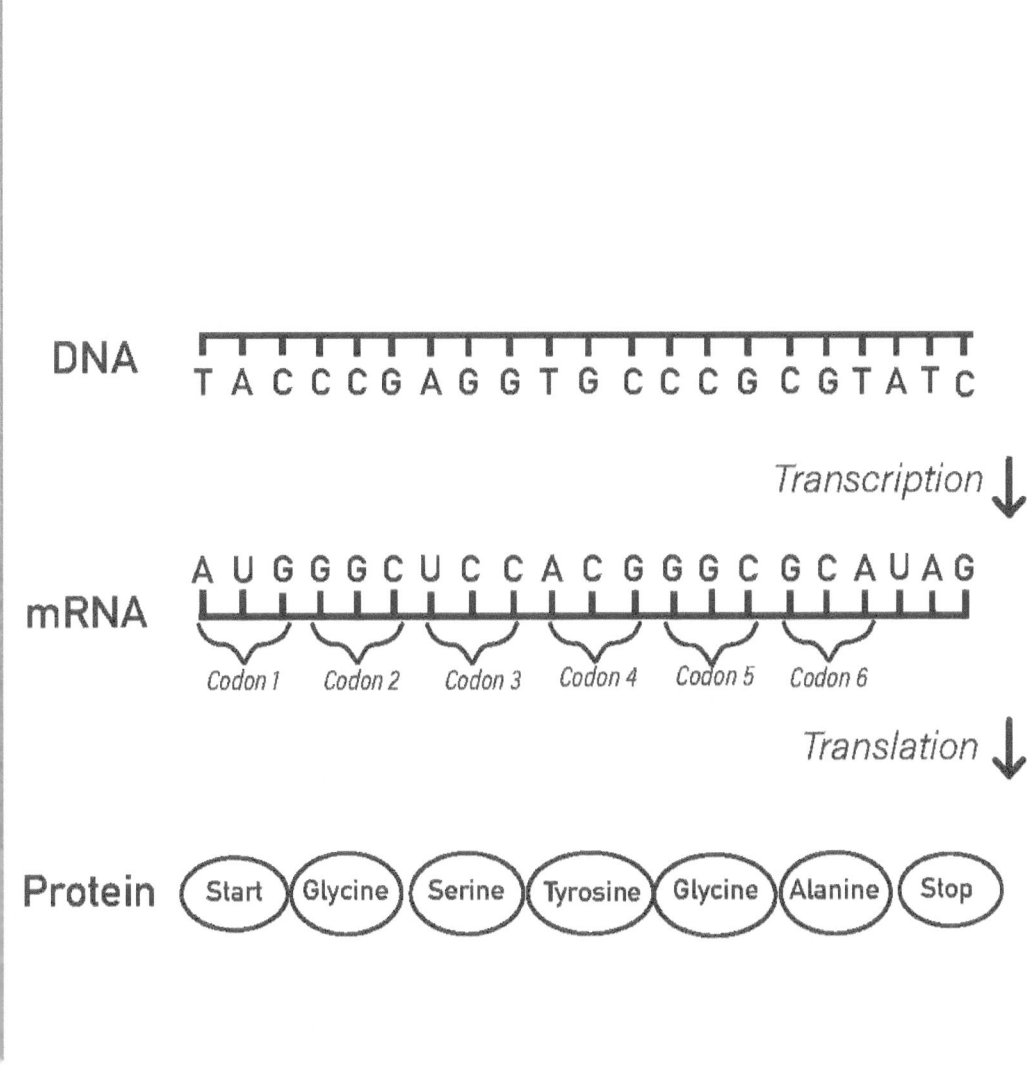

The Genetic Code

In the third and final step, special enzymes mediate a set of complex reactions involving a class of ribonucleic acid molecules called transfer RNAs (tRNAs). These have the ability to ferry amino acids—the building blocks of protein molecules—to their assembly site within the cell. This happens in a sequence specific manner, with the complementary sequence of the respective tRNA aligning with its cognate sequence encoded in the mRNA. This step referred to as translation, is facilitated by the formation of a complex between the mRNA, the amino acid-bound tRNA and a cellular organelle called a ribosome. It is that complex that yields the full complement of proteins produced by cells. The combination of steps one through three is commonly referred to as the Central Dogma of Molecular Biology.

As an undergraduate, I learned about these remarkable developments in the classroom as well as by reading peer-reviewed articles published in a new journal called the *Journal of Molecular Biology (JMB)*. First launched in 1959, *JMB* was a journal that published original, high quality manuscripts authored by some of the most accomplished scientists of the day. I couldn't wait to read the latest findings in what was then my favorite scientific journal.

I was also an avid reader of the *Proceedings of the National Academy of Sciences* (PNAS). That was the official journal of the American National Academy of Sciences. It usually contained articles endorsed by at least one of the Academy's members. I also liked to read manuscripts published in *Biochimica et Biophysica Acta*, an international journal widely followed by my biochemistry and molecular biology colleagues. I did not realize then how

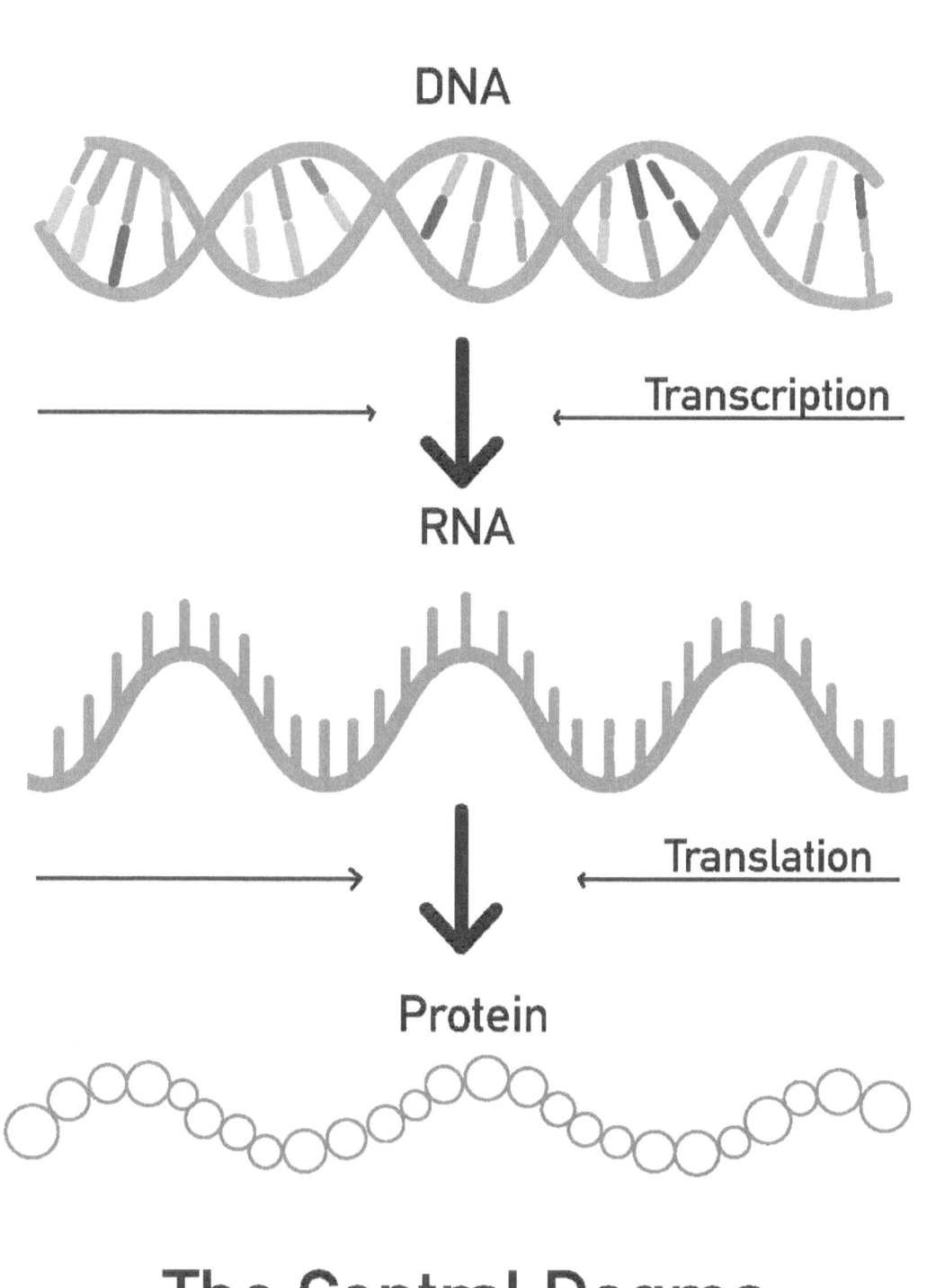

prescient those studies were, but I was sure that something extraordinary was in the making.

Then, Howard Temin, a member of the faculty at the University of Wisconsin, and David Baltimore who was working at the Massachusetts Institute of Technology (MIT), published a report in 1971 showing that cells could be coaxed to make DNA from RNA. Just the thought that the flow of genetic information might be reversible seemed heretical to those who believed in the Central Dogma. Understandably, the Temin and Baltimore manuscript sent shockwaves through the scientific world. They called the enzyme reverse transcriptase.

In the first month on my new job as a research scientist in Big Pharma, I was able to replicate the Temin and Baltimore findings in my own laboratory! I became a believer in what I was reading in my favorite journals.

I was astounded to learn that a tiny RNA-containing virus known as an RNA Tumor Virus (or retrovirus) could insert its RNA genome into a host cell. After penetrating the cell, the retrovirus hijacked the cell's synthetic machinery causing it to make stable viral DNA copies and insert them irreversibly into its own genetic material.

I remember when I learned about a report published by Herbert Boyer, a professor at the University of California in San Francisco, and Stanley Cohen from Stanford University, showing how to splice a foreign gene into a circular DNA molecule known as a plasmid, insert that chimeric construct into a cell, and have the cell express the foreign protein.

The most interesting thing about their work was that it

demonstrated how cells could be used as factories to produce commercially important proteins.

The ability to apply genetic engineering strategies to the mass-production of proteins represented a major breakthrough in applied science. I could sense that classical genetics was rapidly giving way to a new discipline—one that scientists referred to as molecular genetics.

I first met one of the widely celebrated practitioners of molecular genetics in the 1980s. His name is Dr. Oliver Smithies, and at the time of our meeting he was a member of the faculty at the University of Wisconsin. I was a Senior Research Scientist at Pfizer Inc. then, and I had traveled to Madison to learn firsthand about a breakthrough discovery Dr. Smithies made that was advancing the pace and sophistication of genomic science. He showed that embryonic stem cells could be used to introduce specific gene modifications in mice. Armed with such an innovative system he was able to generate transgenic mouse models that mimicked a number of human diseases.

Transgenic animals produced using his technology offered a practical approach for simulating human diseases in a laboratory setting, enabling scientists to use those animal models as reliable targets for the discovery of novel therapeutic agents. Later in my career, I got an opportunity to work collaboratively with Dr. Smithies when we were both faculty members in the University of North Carolina System. Our students and faculty benefited immensely from those interactions.

When Dr. Smithies was on the faculty at the University of Wisconsin, one of his colleagues—Kary Mulis—was credited

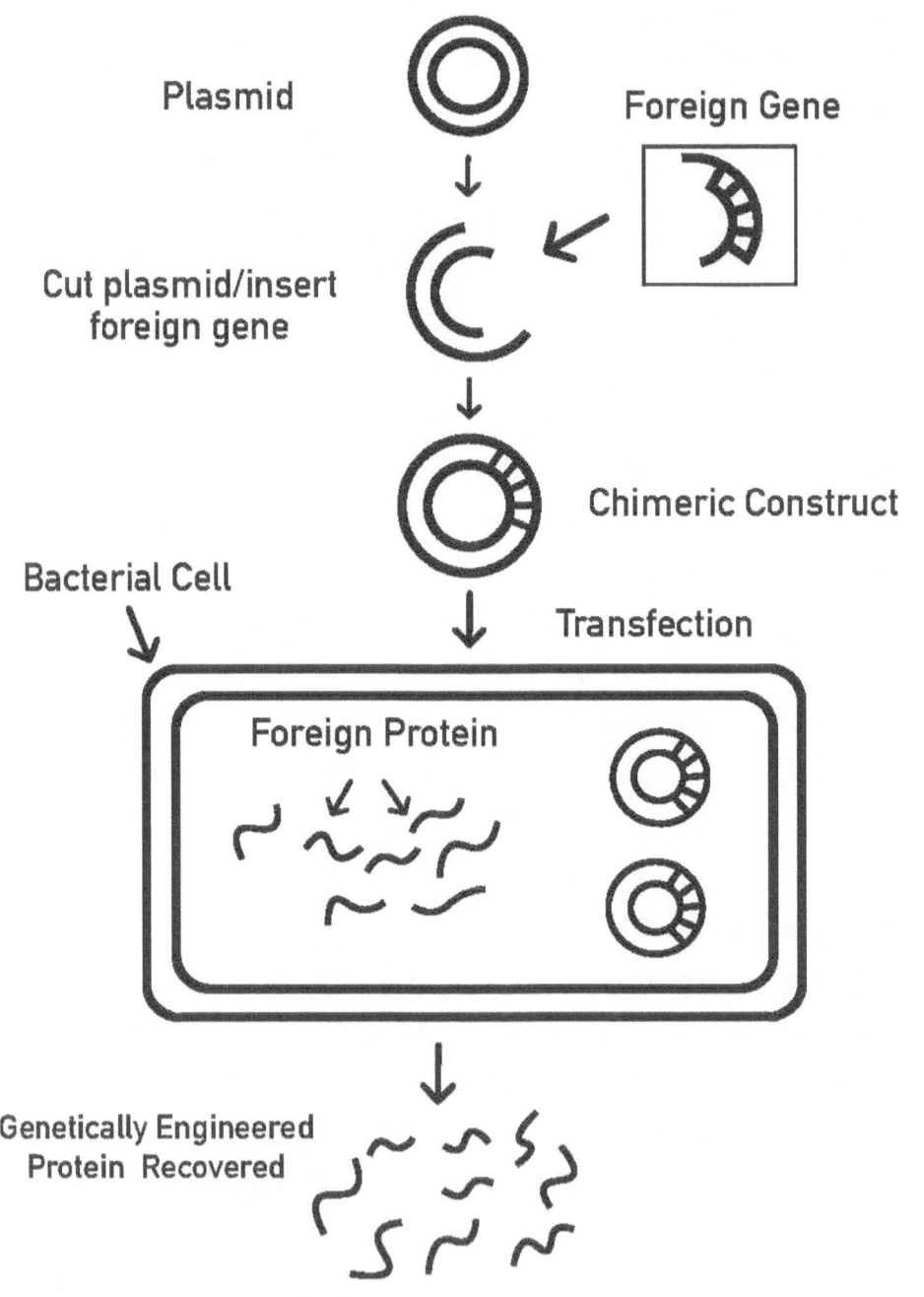

with designing a process called polymerase chain reaction (PCR) that enabled him to mass-produce DNA. PCR accelerated the pace at which specific regions of DNA could be replicated in a laboratory setting for the purpose of studying gene structure and function. Mulis' work ultimately paved the way for the assembly of synthetic genes.

CHAPTER 3

The Biotechnology Revolution

The discovery of restriction endonucleases by Daniel Nathans, Werner Arber, and Hamilton Smith marked the beginning of an era in which biotechnologists were able to manipulate DNA in ways that offered significant commercial advantages. This family of enzymes, isolated from bacterial cells, had the unique ability to bind to the DNA molecule in a sequence specific manner. Nathans, Arber and Smith were awarded the Nobel Prize in 1978 for the discovery of restriction endonucleases.

Prior to the isolation and purification of restriction endonucleases, fragmentation of DNA at specific sites was a major challenge. I remember using a device called a Dounce homogenizer to break DNA strands into pieces so that I could conveniently identify the base at the terminus of each fragment.

The Dounce homogenizer was essentially a glass tube rounded at its base and equipped with a tight-fitting glass pestle. In order to fragment a sample of DNA, I would take a suspension of the molecule and place it in the glass tube. I then inserted the pestle in the tube and gently moved it up and down to break the DNA molecule, generating random fragments. I knew that the size of the fragments produced would vary directly with the number of strokes of the pestle. It was a crude way to fragment DNA that never yielded useful information about the site(s) at which the breaks occurred. The availability of restriction endonucleases rendered the Dounce homogenizer obsolete overnight!

The addition of restriction endonucleases to the molecular biologist's toolkit was a landmark event, and university scientists sought to capitalize on these advances by partnering with venture capitalists. That gave birth to the era of startup biotech companies. Ideas spawned in the laboratories of the nation's research universities were rapidly exported to what was referred to as incubator spaces for more rigorous evaluation. The college professors who were intimately involved in conducting the original research were selected to manage the spin-off companies that resulted from their work.

These innovative businesses that gave birth to the biotechnology industry in the decade of the 1980s were grounded in the conviction that biological systems could be used to develop commercially important products and processes. Utilizing basic concepts from the disciplines of microbiology and biochemistry, molecular biologists devised amazing strategies for manipulating plant and animal genes. The resulting constructs of chimeric DNA were designed to alter metabolic pathways, mass-produce commercially important proteins, and increase the quality of agricultural and veterinary outputs. These revolutionary changes in the large-scale production of biopharmaceuticals expanded America's manufacturing capacity in ways never before imagined.

The rapid pace of development of the new biotechnology industry was also largely dependent on the willingness of venture capitalists to invest in startup companies that were seeking to replace old methodologies with innovative and highly precise new technologies.

Looking back, I can only conclude that those were magical times when almost every startup sought to commercialize a new biopharmaceutical product or develop an innovative cell-based, large-scale production process.

By the end of the 1980s, my recollection is that the U.S. Food and Drug Administration (FDA) approved five proteins for use as biopharmaceuticals. They included insulin, human growth hormone, hepatitis B vaccine, alpha interferon, and tissue plasminogen activator. That number rose to more than one hundred twenty-five genetically engineered drugs by the end of the 1990s.

No one could have ever predicted that there would be approximately one thousand five hundred biotechnology companies in the U.S. by 2006, and that more than three hundred of them would be publicly held. Nothing spoke more directly to the power of innovation and creativity than the incredible pace at which the American biotechnology industry grew in less than two decades.

The rate of new company formation was dramatically accelerated after Congress passed the Bayh-Dole Act. That landmark piece of legislation gave universities, small businesses and non-profit organizations greater control over their own inventions. One significant outcome of the Bayh-Dole Act was an exponential increase in the number of patents and royalties awarded to American scientists.

I joined the Association of University Technology Managers (AUTM) in the 1990s in order to keep pace with the rapidly changing landscape of intellectual property ownership and protection. Membership in AUTM enabled me to become acutely aware of the role that venture capitalists played in the wave of new company formation that was sweeping across America. I met with the founders of many start-ups whose innovative strategies were yielding windfall profits for investors.

CHAPTER 4

The Human Genome Sequence

The decade of the 1980s brought a new set of challenges to the fledgling biotechnology industry. It involved a call to use existing tools for DNA manipulation to sequence the entire human genome.

James Watson wrote a report addressing the feasibility of initiating such a project. His report, combined with the advances highlighted above and buttressed by phenomenal improvements in automated DNA sequencing, paved the way for what became an extraordinary call to launch the Human Genome Project.

It seemed like such a daunting task when it was first enunciated that it elicited modest support from some of Watson's colleagues. As time went by however, and with the emergence of new tools and technologies, the probability of successfully prosecuting a project of that scale appeared to be more realistic.

In 1987, the Department of Energy (DOE) led the way by initiating a study designed to better understand the mutagenic effects of radiation on the human genome. I followed developments on that project closely, and when a group of prominent biologists advocated for a linkage between the DOE initiative and the idea of sequencing the human genome, many scientists felt that their proposal was unrealistic. I agreed with them based on my own experience as well as on the state of DNA sequencing technology in 1987. I knew that automated sequencing was available, but in the late 1980s the technology was costly, error prone, and sequencing speeds were unimpressive. I could not envision how it would be possible for even the most accomplished investigators to successfully apply the existing machines to the analysis of a genome that was estimated to be more than three billion base pairs in size. Additionally, there was the real concern of deciding how such a project would be managed. Questions were raised about leadership, strategic planning, data collection and analysis, and funding mechanism(s). Some of those questions were answered in 1988 when Congress authorized NIH to support an effort to explore the concept of analyzing the entire Human Genome. The agency used the funds that Congress made available to establish an Office of Human Genome Research, and selected James Watson as its director. I remember how politically charged the atmosphere was when the decision was announced. Many scientists, as well as the institutions they represented, actively competed for the right to participate in the implementation phase of the project.

After assuming responsibility for directing the Office of Human Genome Research, Watson was pushed to the limit. One year after he arrived at the NIH, the name of the agency he headed was changed to the National Center for Human Genome Research (NCHGR).

Because of his larger-than-life reputation, Watson seemed to be a good fit to lead the NCHGR. After all, his was the face of the biotechnology revolution. Like many of my colleagues, I felt that he had the track record needed to get things done. He proved that he was up to the task when in 1990 he issued a joint research plan to fund the first five years of the Human Genome Project. I could not have predicted then that two years later he would demit office and be replaced by Dr. Francis Collins.

Despite the air of uncertainty following Watson's resignation from the NCHGR, the NIH effort remained on track. Under the direction of its new leader, the NCHGR assembled a consortium comprising some of the most talented investigators in the field. That group was tasked with cataloging and analyzing massive amounts of data generated by multiple DNA sequencing centers. Then in 1997, the Department of Health and Human Services elevated the NCHGR from a center to an institute, renaming it the National Human Genome Research Institute (NHGRI), and retaining Francis Collins as its director. The effort of the NHGRI ultimately succeeded, yielding the first draft sequence of the human genome much sooner than had been initially envisioned. In fact, the draft sequence of the entire human genome was published in the February 2001 issues of the journals *Science* and *Nature*.

As I perused the cover stories of those journals, I could not think of any project that showcased American creativity, ingenuity, and innovation like the Human Genome Project. I had no doubt that the draft sequence would augur well for the development of a variety of innovative strategies to treat communicable and non-communicable diseases. It promised a wave of new developments, ranging from the use of plants to manufacture novel biopharmaceuticals, to designing rational approaches for producing personalized medicines.

The mere thought of having the ability to develop designer drugs was sufficient to galvanize significant public support for the new industry, and this was evidenced by an overwhelmingly positive response from the investor community.

There is no doubt that the sequencing of the human genome in 2001 represented yet another example of America's global competitiveness in science and technology. I was proud of the fact that I knew many of the scientists who participated in its planning and execution, and I was thrilled to have had the opportunity to meet the two American scientists who played major roles in guiding the project to its successful completion.

After publication of the draft sequence, Francis Collins was a special guest at a meeting held at the University of North Carolina at Chapel Hill (UNC-CH) to discuss the ethical, legal, and social implications (ELSI) of genetic testing. ELSI was a major concern then, and many Americans worried about the possible misuse of the technology to disadvantage members of ethnic and racial minority groups. I found Dr. Collins to be personable and unpretentious. None of those in attendance at that meeting ever doubted his competence or sincerity. I learned that he received his training in medicine from UNC-CH in 1977. Prior to that, he had earned a doctoral degree in physical chemistry from Yale University (1974), and in 1981 he returned to Yale as a Fellow in the department of Human Genetics. He worked in the laboratory of Dr. Sherman

Weissman, an outstanding scientist who collaborated with one of my colleagues at the Central Research Division of Pfizer.

I also met Craig Venter, the founder of Celera Genomics and The Institute for Genomic Research (TIGR) and head of the private sector initiative to sequence the Human Genome. His was an amazing story. I read his book long after I met him. In that account about his life which he entitled *A Life Decoded*, he was forthright in stating that his "early years were hardly a model of focus, discipline and direction."

After serving in Vietnam, he began his academic career at the College of San Mateo in California with support from the GI Bill. When he graduated from community college, he enrolled in the University of California San Diego (UCSD) where he completed masters and doctoral degrees. In a surprising move, Venter decided that he would forego seeking a position as a postdoctoral trainee. That was not the traditional career path for a research scientist, but he was undeterred in his desire to obtain a faculty position at a medical research center. His big break came when he was offered a faculty position at the Roswell Park Cancer Institute at the State University of New York at Buffalo. In 1983, he moved from Roswell Park to NIH where he joined the intramural program at the National Institute of Neurological Disorders and Stroke.

I first learned of Venter's interest in constructing complimentary DNA (cDNA) libraries when I worked at the Central Research Division of Pfizer. I did not meet him then, but I got that opportunity when I was a faculty member in the University of North Carolina System. He gave an incredible presentation to a large group of scientists, administrators and elected officials when he visited the Research Triangle Park to talk about his work on the Human Genome Project. I must admit that what intrigued me most was his avowed interest in creating artificial life.

I still marvel at the accomplishment of Collins and Venter. Their career paths were very different, their personalities seemed to be poles apart, they embraced different sequencing strategies, and yet their quest for greatness was ultimately attained in 2001. I admire their ingenuity and pragmatism, and heartily commend them for their ability to successfully complete a project as complex as the sequencing of the human genome.

Just like James Watson, Francis Collins and Craig Venter represent modern day examples of great American inventors of the nineteenth and twentieth centuries. The successes of these individuals speak clearly to their creativity, incredible management skills, and unwavering commitment to the business of science.

CHAPTER 5
America's Global Leadership in Science

Completion of the sequencing of the human genome in 2001 is another example of America's preeminence in the areas of science and technology. Americans planned the project and played a significant role in its execution. Partnership initiatives between public and private sector groups served as its primary funding mechanism, and two Americans, Francis Collins and Craig Venter, provided extraordinary leadership. At every step along the way, American ingenuity was a key contributor to the project's timely execution.

For generations, Americans have drawn on their creative energies to develop some of the world's most useful products and processes. During the eighteenth and nineteenth centuries, Europeans were intensely focused on understanding the basic principles of science. Textbooks were filled with examples of their contributions to the formulation of the laws and theories of biology, chemistry and physics. American pragmatists were taking a different approach—they preferred to invest inordinate amounts of time and money in trying to convert theory into practice.

That emphasis on applied science enabled Americans like Eli Whitney, Thomas Edison, the Wright Brothers, Henry Ford and George Washington Carver to couple Yankee ingenuity with advances in theoretical science. The result was the invention of the cotton gin, the light bulb, the first successful airplane, the automobile, and a variety of new agricultural crops.

America's dominance in the applied sciences continued throughout the twentieth century, as the nation recorded spectacular advances in aeronautics, automobile production, transistor technology, integrated circuitry, computer science, and telecommunications.

Among the many successes credited to American ingenuity during that period of intense technological innovation, none provided a greater competitive advantage than the effort to harness nuclear energy. Drawing on advances resulting from splitting the atom, American scientists created the world's first atomic bomb. Unquestionably, acquisition of nuclear weapons was singularly responsible for elevating American military power to an unrivaled position globally.

Next there was the space age. In response to the launching of the world's first artificial satellite (Sputnik 1) by the Soviet Union in 1957, American scientists drew on advances in rocketry, material sciences, and computer technology to develop the Apollo program. That initiative involving human spaceflight was funded by the National Aeronautics and Space Administration (NASA) in 1961, and resulted in American astronauts landing on the surface of the moon by 1969.

Five subsequent Apollo missions placed more American

astronauts on the moon, an accomplishment that to date has not been replicated by any other nation.

America's six lunar landings have been considered to be the greatest technological achievement ever. The images of American astronauts walking on the moon are still etched in the minds of many, providing incontrovertible proof of America's global dominance in science and technology.

There are countless other examples of major projects undertaken by American scientists that have been incredibly successful. That legacy has been widely admired and emulated by business leaders all across the world.

Many have often wondered why America has been so successful in outstripping its competitors in the applied sciences. They ask how can a nation whose students perform so poorly in science, technology, engineering, and mathematics (STEM) disciplines rank so highly in patent output and new product/process development? One way of responding to this query is to attribute America's successes in science and technology to its ongoing practice of importing talent. While this might have some basis in fact, there are still those who believe that America's achievements in science result from its enduring commitment to fostering a competitive business culture that emphasizes investing in basic research and rewarding translational outcomes.

I learned that firsthand when I joined Pfizer in 1971. I found out then that Big Pharma was one of the primary beneficiaries of advances made during the twentieth century. Many flagship companies in that industry got their start in the chemistry business, and as circumstances changed, they broadened their portfolios to include antibiotics and other human healthcare products. For example, an accidental discovery by a British scientist, combined with devastating losses to Allied troops during World War II, created a scenario that propelled the fledgling American chemical industry into the human healthcare arena.

That shift in business strategy began when British scientist Sir Alexander Fleming noticed that one of the bacterial cultures he was studying was contaminated by a fungus. He later observed that bacterial colonies immediately surrounding the fungus were being destroyed. After further investigation, he realized that the fungus was producing a substance that inhibited the growth of bacterial cells. Fleming called that substance penicillin, and his insightful observation marked the beginning of the age of antibiotics.

Although penicillin was discovered in 1929, it was not until 1940 that Dr. Howard Florey, a colleague of Fleming's, realized how important it would be to mass-produce penicillin to treat bacterial diseases that were devastating Allied forces during World War II.

The prospect of using the new wonder drug to reduce troop morbidity and mortality did not go unnoticed by the American government. In 1943, the U.S. War Production Board drew up a plan for mass-producing penicillin. When the major pharmaceutical companies were asked to assist with that effort, twenty-one of them were selected to participate in the penicillin program. Lederle, Abbott Laboratories, Squibb, Merck and Pfizer were among the companies chosen.

At that time, Pfizer, a Brooklyn-based chemical company,

was widely known to be a major manufacturer of citric acid. Taking advantage of its extensive experience with fermentation, the company embarked on a production strategy that utilized its novel, deep-tank process. Yields of penicillin were so impressive that Pfizer became the world's largest producer of that "miracle drug."

The success of penicillin convinced many American chemical companies to place greater emphasis on discovering novel antibiotics, and several of them expanded their R&D portfolios to include funding projects on diseases related to human and animal health.

I've heard incredible stories about these companies undertaking massive soil screening campaigns in a relentless search for the next "million dollar bug." Every microbial culture identified through such efforts was grown under specially controlled laboratory conditions to determine whether it had the capacity to produce a novel antibiotic. Whenever a putative candidate was identified, medicinal chemists would analyze its chemical structure, attach sub-groups to modify its architecture, and test the product for biological activity. This method commonly referred to as structure-activity relationship or SAR, enabled medicinal chemists to select the analog displaying the highest level of bioactivity and the least toxicity as the candidate for development.

It is not surprising to learn that the phenomenal growth of the American pharmaceutical industry occurred when antibiotics were considered to be more attractive to manufacture than bulk chemicals.

As the fledgling industry grew, some companies worked on

designing novel drugs to treat chronic diseases such as diabetes, arthritis, and cancer. That shift from a high-volume-low-value production process to one that was referred to as high-value-low-volume, paid enormous dividends. In fact, it was not unusual for companies to launch several blockbuster drugs simultaneously, some with sales exceeding a billion dollars annually.

In 1991, I distinctly remember reading an article about manufacturing competitiveness in the business section of a leading U.S. weekly magazine. That report cited five U.S. firms that were "beating up foreign competitors as they slugged their way to the top of the global marketplace." One of those companies was Pfizer Inc. that at the time was recording numerous marketplace victories. According to the magazine, Pfizer's success was believed to result from strategically targeted R&D spending. The company's ranking then, based on global drug sales, was eleventh in the world. What is remarkable about that story is that by the end of the twentieth century, Pfizer would become the number one pharmaceutical company in the world!

Today, although the U.S. biopharmaceutical sector accounts for the largest share of the nation's R&D activities, the industry as a whole is still facing considerable uncertainty. Some of this may be due to the way globalization is transforming the manufacturing business. Most of it however, is driven by consolidation of major players through mergers and acquisitions, patent expirations, increases in competition from the large generic companies, and growing uncertainty about America's healthcare policy.

Based on industry trends and outlooks, analysts think that over the next decade, Big Pharma might be forced to shift from a mass-market to a target-market approach in order to increase revenues. Any change in the marketing model could have an adverse impact on the number of employees hired by the sales and marketing divisions of these vitally important American companies.

In the post-blockbuster drug era, Big Pharma will also be under significant pressure to transition from the old paradigm of developing exclusive, patented drugs to a newer model that many in the industry are still trying to define. Some companies are exploring the possibility of abandoning the traditional R&D approach and embracing a virtual mode of operating with multiple players involved at each phase of the drug discovery process.

The addition of biopharmaceuticals to Big Pharma's portfolio of products has increased R&D expenditures dramatically, and the returns on those investments have been phenomenal! In order to maintain the successes recorded by this critical sector of the American economy, Big Pharma will have no choice but to draw on the creative energies of its most gifted and talented employees. The possibilities of meeting this challenge in the post-genomics era seem limitless.

In the new millennium, regardless of whatever organizational structure Big Pharma assumes, the ability to maintain and extend its leadership position in science and technology will depend on the extent to which America's colleges and universities deliver on their primary mission of educating and training future generations of basic and applied scientists.

CHAPTER 6

The American Education System

The American education system is made up of a mixture of primary, secondary and tertiary institutions, each dedicated to preparing students to be productive citizens. Control of this complex assortment of organizations rests with the states, and is delegated to school boards, institutional governing bodies, and state secretaries of education. DoED is the agency responsible for providing guidance and support to the American education system at the federal level.

Students typically enter the education pipeline in kindergarten. They transition through primary, middle and high school, and are awarded diplomas after successfully completing requirements at each stage of the public school system.

The average high school graduate has the option of signing up with a state-supported community college to earn an associate's degree, or enrolling in a four-year college to pursue a bachelor's degree.

College graduates receiving the highest grade point average and standardized test score qualify for admission to graduate schools. These institutions offer masters and Ph.D. degrees in highly specialized areas. On average, it takes a masters degree student two years to graduate, while doctoral degree candidates spend at least four years before satisfying all degree requirements.

The development of America's public schools and higher education system dates back to colonial times. Many of the early settlers patterned their institutions after two British flagship universities—Oxford and Cambridge.

Altogether nine institutions of higher learning were established in the original thirteen colonies. Harvard College, founded in 1636 by the Massachusetts Bay colony, was the oldest. The College of William and Mary in Williamsburg, Virginia, chartered in 1693, was the second oldest, and the third oldest, Yale University in New Haven, Connecticut, opened its doors in 1701.

Other colleges and universities launched prior to independence include Princeton University in Princeton, New Jersey (1746); Columbia University in New York City (1754); the University of Pennsylvania in Philadelphia, Pennsylvania (1755); Brown University in Providence, Rhode Island (1764); Rutgers University in New Brunswick, New Jersey (1766); and Dartmouth College in Hanover, New Hampshire (1769).

Similar colleges modeled after British universities were established in the south, as the American higher education system continued its expansion well into the eighteenth century.

In 1775 the colonists, upset by British rule protested vehemently, and their public outcry escalated into overt war. That struggle for freedom culminated with the declaration of independence in 1776, seven years before hostilities ended in 1783.

The nineteenth century was a period of intense conflict between northern and southern states. The sudden, forced importation of large numbers of Blacks from Africa in the seventeenth century created an untenable situation within colonial America that brought southern plantation owners into direct conflict with Christian morality.

Not surprisingly, the forces of capitalism prevailed, creating a system of slavery within the colonies, provinces, and territories that inflicted considerable pain and suffering on enslaved persons, their families, and the communities in which they were forced to live and work. These practices ultimately led to a major rift between northern and southern states, triggering a brutal civil war. The Union army eventually prevailed over Confederate forces, and emancipation was officially declared in 1863.

Prior to emancipation, two normal schools, Cheyney and Lincoln, were chartered in Pennsylvania to educate the children of enslaved persons. Institutions established for that particular purpose have been referred to as HBCUs.

Cheney is the older of the two normal schools established in Pennsylvania. It was founded in 1837 by a white American philanthropist. Its benefactor, Richard Humphreys, was born in the West Indies and was thought to be a member of the Quaker movement. It is reported that he donated ten thousand dollars to establish the Institute for Colored Youth in Philadelphia. In

1902 that school was relocated to George Cheyney's farm west of Philadelphia, and in 1913 its name was changed to Cheyney State Teacher's College. In 1983, after becoming a member of the state system of higher education, Cheyney State Teacher's College was renamed Cheyney University of Pennsylvania. Today, Cheyney is situated on a two hundred seventy-five acre campus in suburban Pennsylvania. Its enrollment is estimated to exceed one thousand students.

Lincoln is the other normal school established in Pennsylvania and the first degree-granting HBCU in America. Lincoln, located near the town of Oxford, got its start in 1854, nine years before President Abraham Lincoln signed the Emancipation Proclamation. Its founder was John Dickey, a Presbyterian minister who named it Ashmun University. His purpose was to use that institution to educate male youth of African descent. The school was later renamed Lincoln University after President Lincoln was assassinated.

During the first one hundred years of its existence, Lincoln University educated twenty percent of black physicians and ten percent of black lawyers. The famous civil rights lawyer and Supreme Court Justice Thurgood Marshall was a graduate of Lincoln University. Other famous Lincoln University alumni include the internationally acclaimed African American poet Langston Hughes, and the first president of Ghana, Dr. Kwame Nkrumah. In 1972 Lincoln University's status was changed to that of a state-related institution within the Commonwealth of Pennsylvania System.

The trend of creating colleges to educate black and white Americans separately continued during the nineteenth century when the higher education system underwent its most significant expansion. The vehicle for that change was legislation proposed by Congressman Justin S. Morrill from Vermont.

The first Morrill Act, approved by Congress in 1862, was intended to establish the land-grant colleges—institutions dedicated to developing programs in agriculture, as well as military and mechanical arts. Because the law was enacted one year after the Civil War began, participation in the land-grant program was restricted to Union states exclusively.

Provisions of the first Morrill Act included a grant of thirty thousand acres of federal land to each qualifying state. Eligibility for the grant was based on the number of seats apportioned by Congress in the 1860 census. States receiving the award were extended the option of constructing at least one college on the land, or using proceeds from its sale to establish land-grant institutions elsewhere. More than seventeen million acres of land were allocated by the federal government to states involved in the land-grant program.

Twenty-eight years later, Congress approved the second Morrill Act. That legislation authorized annual appropriations to each participating state, and permitted former Confederate states to benefit from the program. One of the provisions of the revised law was that it prevented states receiving Morrill Act funds from refusing to admit Blacks to their land-grant colleges. In order to circumvent that requirement, Congress allowed states opposed to integration to create separate institutions to meet the educational needs of Blacks. That concession by Congress, embedded in the second Morrill Act, tacitly endorsed what eventually became a collection of racially segregated land-grant colleges.

Several HBCUs benefited from passage of the second Morrill Act. Among those were Kentucky State University (KSU) in Frankfort, Kentucky; Florida Agricultural and Mechanical University (FAMU) in Tallahassee, Florida; North Carolina Agricultural and Technical State University (NC A&T) in Greensboro, North Carolina; and Tuskegee University in Tuskegee, Alabama.

KSU is a public HBCU that was founded in 1886 as the State Normal School for Colored Persons. In 1890, when it became a land-grant college, departments of home economics, agriculture and mechanics were added to its academic program. In 1902, the name was changed to Kentucky Normal and Industrial Institute for Colored Persons. Subsequent name changes occurred in 1926 (Kentucky State Industrial College for Colored Persons); 1938 (Kentucky State College for Negroes); and 1952 (Kentucky State College). In 1972 Kentucky State College was elevated to a university. Today, KSU occupies an eight hundred eighty-two acre campus that includes an agricultural research farm and an environmental education center. It offers degrees at the associate, undergraduate and post-graduate levels to more than two thousand students.

FAMU was founded in 1887 as the State Normal College for Colored Students. It became a land-grant college in 1891 when its name was changed to the State Normal and Industrial College for Colored Students. In 1909, it underwent another name change becoming the Florida Agricultural and Mechanical College. In 1953, it was renamed Florida Agricultural and Mechanical University.

FAMU's greatest achievement came during the period between 1950 and 1968 when it created the schools of Graduate Studies, Law, Nursing, and Pharmacy. A hospital, established on its campus in 1956, served as the only medical facility for Blacks within one hundred-fifty miles of the state capital in Tallahassee.

FAMU made remarkable strides in the last decade of the twentieth century under the leadership of its president Dr. Frederick Humphries. During that period, the university was credited with recruiting more National Achievement Scholars than Harvard University. It was also selected as College of the Year by *TIME Magazine, Princeton Review*, and cited by what was then known as *Black Issues in Higher Education* as the institution that awarded more baccalaureate degrees to African Americans than any college or university in the nation.

NC A&T is an HBCU located in Greensboro, North Carolina. It is one of sixteen constituent institutions in the University of North Carolina System. It was established in 1891 by an act of the North Carolina General Assembly to take advantage of the provisions of the second Morrill Act. The name given to the institution at that time was North Carolina's Agricultural and Mechanical College for the Colored Race. In 1915, the state legislature changed its name to the Agricultural and Technical

College of North Carolina. And in 1967, when the college was elevated to university status, the name was changed to NC A&T.

NC A&T is currently the largest HBCU in America with an enrollment of more than twelve thousand students. The university's College of Engineering ranks first in the nation, based on the number of undergraduate degrees awarded to African Americans. It is also recognized as the leading producer of African American engineers at the masters and doctoral levels.

Tuskegee University is an HBCU located in the city of Tuskegee, Alabama. It was established in 1881 as the Normal School for Colored Teachers. Its founder Lewis Adams was an African American and former enslaved person. This internationally known institution has the distinction of having the noted black educator Booker T. Washington as its first president.

During president Washington's tenure, which lasted from 1881 to 1915, the school made remarkable strides, becoming a national leader for training Blacks in the technical and vocational areas. Among president Washington's most notable achievements was his ability to gain land-grant status for Tuskegee in 1892. With state support provided through the second Morrill Act, Tuskegee expanded its curriculum offerings and hired new faculty.

In 1896, the legendary African American scientist Dr. George Washington Carver was hired to head the Department of Agriculture. His research on crops such as peanuts, soybeans and sweet potatoes elevated Tuskegee's reputation both nationally and internationally.

Tuskegee is also widely recognized for its role in establishing the Tuskegee Home for African American Soldiers. That facility, built on land donated by the university, was dedicated to providing long-term care for black veterans returning from World War I. It was brought into the Veterans Administration Hospital System in 1930.

Another of Tuskegee's most notable achievements was the creation of the Army Air Corps in 1941. That program is credited with training the Tuskegee Airmen who became one of America's most respected fighter groups of World War II.

The Tuskegee School of Veterinary Medicine was launched in 1945, becoming the first such program at an HBCU. It has an impressive record of achievement, having trained seventy-five percent of all of African American veterinarians.

Whether created by colonists, Quakers, former slaves, an act of Congress, or other means, the American education System has been remarkably resilient. It has survived a revolutionary war, a Civil War, and remained intact in the face of one of its greatest challenges—the fight to end racial discrimination during the twentieth century.

CHAPTER 7

Integrating American Higher Education

Today there are approximately one hundred seven HBCUs in the U.S. serving more than two hundred fifty thousand students. Fifty-one of them are public institutions and fifty-six are private. Consistent with their mission, HBCUs continue to award thousands of baccalaureate degrees annually to African American students. Their graduate and professional schools routinely train Blacks in medicine, dentistry, pharmacy, law, criminal justice, architecture, business, and a number of other disciplines. In spite of these positive contributions to America's higher education system, there are still voices in state legislatures that constantly seek to roll back the gains made by the nation's HBCUs.

Such calls to devalue the status of state-supported HBCUs are not uncommon, and represent stark reminders of a distant past. What is surprising about these attempts is that they seek to use the legislative process to constrain growth and expansion of HBCUs. Ironically, when Congress passed the second Morrill Act, the intention was to create a separate-but-equal framework permitting Blacks to run their own institutions.

While the Morrill legislation authorized the establishment of land-grant colleges to serve an expanding black population, it offered states the option of maintaining separate facilities for Blacks and Whites. The resulting funding inequities, compounded by overt acts of racial discrimination, caused black Americans to seek remedy through the courts. Although these challenges to Jim Crow laws were not directly initiated by HBCUs, they would serve to bring an end to one of the most outrageous practices engaged in by a nation that desperately needed to educate all of its citizens.

The most noteworthy of these challenges to racial segregation occurred in 1892 when a black man named Homer Plessy was arrested on a train in New Orleans for refusing to give up his seat to a white man. Plessy claimed that the Louisiana law permitting the separation of Blacks from Whites violated the equal protection clause of the Fourteenth Amendment to the Constitution, and he filed a case with the courts. Plessy's case reached the Supreme Court in 1896, and the justices ruled that as long as the separate facilities for Blacks and Whites were equal, segregation did not violate the Constitution.

In 1938, a black man named Lloyd Gaines filed a case against the University of Missouri Law School claiming that he was denied admission because of his color. The Supreme Court ruled that in order for the school to be in compliance with the equal protection requirement of the law, the state of Missouri would have to provide training for every qualified citizen. The Court further stated that if Missouri failed to provide Mr. Gaines with access to a separate-but-equal law school, it would be obligated to admit him to its white law school.

Following the Gaines decision, governors of southern states responded by permitting a select number of state supported HBCUs to offer professional courses in law, pharmacy, and medicine.

It was not until 1954 that the Supreme Court, in a landmark decision, declared separate-but-equal to be unconstitutional. The case, known as Brown versus the Board of Education of Topeka, Kansas, was a consolidation of five separate cases heard by the Court.

One of the plaintiffs was a black man named Oliver Brown. He brought charges against the Board of Education of Topeka, Kansas concerning the issue of state-sponsored segregation in public schools. Thurgood Marshall, who later became a justice on the Supreme Court, presented the case on behalf of the National Association for the Advancement of Colored People (NAACP) Legal Defense Fund. Marshall argued that having separate schools for Blacks and Whites was inherently unequal and violated the Fourteenth Amendment.

In May of 1954, the Warren Court ruled unanimously that state laws establishing separate public schools for Blacks and Whites were denying black students equal educational opportunities. It took several additional hearings, and the cooperation of attorneys general from all states guilty of enacting laws permitting segregation in public schools, before plans were submitted indicating how to proceed with desegregation.

The Supreme Court's decision dealt a major blow to legal

segregation in American educational institutions, but did little to improve the plight of Blacks seeking admission to the nation's more prestigious institutions.

The Civil Rights Act of 1964 eventually outlawed racial segregation in all schools, public facilities, and places of employment. Title IV of that Act encouraged the desegregation of public schools, and authorized the U.S. attorney general to file lawsuits to enforce its provisions.

In spite of these very positive steps taken by the federal government to end the injustices perpetrated against Blacks by states, the decade of the 1960s was one of the most turbulent in American history. During that period, proponents and opponents of racial segregation battled each other on the streets in the north and in the south. Nightly television news reports were replete with images of civil rights groups and freedom riders campaigning for their rights, and angry white mobs, baton-wielding police, vicious dogs, and white officials either attacking the marchers or blocking the entrances to their segregated schools and universities. National Guard troops were often sent in by the federal government to quell these disturbances.

The nation was torn apart by church and school bombings, the brutal murders of civil rights workers in Mississippi, the assassination of President John Kennedy and his brother Robert Kennedy, as well as civil rights leaders Medgar Evers, Dr. Martin Luther King Jr. and Malcolm X.

Regardless of these brutal confrontations with groups opposed to integration, the civil rights movement, aided by unwavering support from the NAACP, the National Urban League, the

Congress of Racial Equality and other organizations, was relentless in its quest for desegregation.

Finally, as the decade of the 1970s began, the barriers to segregation started to fall across the nation and integration of America's educational institutions became the order of the day.

One of the major consequences of integration was that white institutions initiated a campaign to recruit top-tier faculty and administrators from HBCUs. While those initiatives opened doors that were previously closed to Blacks, they dealt a devastating blow to HBCUs. Several black universities were forced to close law schools, medical schools, and even hospitals. It was not uncommon during that period to find the names of some of the most accomplished African American educators prominently displayed on the rosters of the nation's Ivy League schools. I have often heard African American professors refer to that period as the time when the "A" team at HBCUs was replaced by their "B" team.

As the call for greater diversity reverberated across the American higher education landscape, the trend of recruiting talented faculty from HBCUs continued. The result was that most of the members of the "B" team were lured away from HBCUs to fill faculty slots at middle-tier white institutions. That hemorrhaging of highly talented faculty from minority serving institutions did little to help them meet the increasing demand to train the next generation of African American students.

Inadequate funding by state legislatures was another reason

for the slow pace of growth of many state-supported HBCUs. It was not unusual for state funding formulas to favor majority institutions over HBCUs. Faculty compensation was another area in which states favored majority over minority institutions.

One of the special privileges states extended to majority institutions was the practice of allowing them to admit only the brightest and best high school graduates to their freshman classes. This practice enabled majority institutions to consistently report higher graduation rates than their HBCU counterparts.

White institutions were also able to supplement their operating budgets with endowments from alumni and wealthy philanthropists. HBCUs on the other hand, were not the beneficiaries of such largess, and as a consequence have seen their fortunes rise or fall as appropriations from state legislatures increased or decreased. Racial attitudes perpetuated by the popular belief that Blacks were inferior to Whites, made it easier for states to justify those horrible practices.

Limited by the size of their physical plants, operating budgets, and endowments, many HBCUs have worried when their programs, services, and operations were subjected to review by external accrediting agencies. That task was assigned to private educational associations approved by the DoED.

One privately managed agency interested in promoting academic excellence at America's higher education institutions was the Carnegie Classification System of Colleges and Universities. Some aspects of how that classification system worked are presented in the next chapter.

CHAPTER 8

Classifying Colleges and Universities

As indicated earlier, the Morrill Act(s) played a significant role in strengthening and expanding postsecondary education in America. After the major struggles of the 1960s, the American higher education system increased in both size and diversity. By the end of the 1970s, there were over three thousand five hundred colleges and universities in America serving more than twenty million students through an eclectic mix of state, regional, religious, military, racial and tribal institutions.

Because of its size and complexity, this extraordinarily diverse collection of educational institutions literally operated on an ad hoc basis. This was primarily due to the fact that the federal government's role in education is limited by the Tenth Amendment to the Constitution.

Congress, recognizing that the best way to ensure that funding mechanisms were administered fairly and reliably, sought to establish a framework for classifying these institutions. There was general agreement that criteria used to achieve the desired level of objectivity should include measures such as quality and quantity of graduate program offerings; type and number of degrees awarded; fields of specialization; and amount and sources of federal research funding awarded.

The Carnegie Foundation for the Advancement of Teaching, an independent policy and research center chartered by Congress in 1906, was the organization that was assigned the task of developing a classification system for all of the nation's colleges and universities. And it authorized its Commission on Higher Education to implement the program.

I spent a considerable amount of time familiarizing myself with the Carnegie Classification System when I served as Professor of Biology at City College of the City University of New York in the early 1990s. I learned then that the original framework for the Carnegie Classification System was drafted in 1970 and published in 1973 by the Carnegie Commission on Higher Education. It categorized U.S. colleges and universities using data collected from surveys administered by the Integrated Postsecondary Education Data System (IPEDS)—the core postsecondary data collection program of the U.S. National Center for Education Statistics (NCES).

IPEDS conducted surveys annually and made them available to the Carnegie Commission. In order to be eligible to participate in the survey, American colleges, universities, technical and vocational schools were all required to demonstrate that they were fully compliant with criteria established by Congress through the DoED, and certified by a group of private, nationally recognized organizations known as accrediting agencies.

In its basic scheme, the Carnegie Classification System

identified six classification groups: doctoral granting universities; masters colleges and universities; baccalaureate colleges; associates colleges; specific focus institutions; and tribal colleges. Highlights of criteria used to define each of the six classification groups are provided below.

The doctoral granting group included institutions classified on the basis of level of research, research expenditure, number of doctoral degrees awarded, and number of research faculty employed. These were sub-divided into three categories—institutions with very high research activity; institutions with high research activity; and doctoral and professional universities.

Doctoral research universities with very high research activity were considered to be the nation's premier higher education institutions. This was based on annual output of peer-reviewed publications; the number of investigator initiated grants and patents awarded; the number of Ph.D. degree candidates graduated, and performance in several other scholarly areas.

Doctoral research universities designated as institutions with high research activity, and those assigned to the doctoral/professional university group were placed in those categories based on their respective outputs in each of the performance areas listed above for doctoral research universities with very high research activity.

Masters colleges and universities were categorized as institutions that awarded at least fifty masters degrees, but fewer than twenty doctorates. Those fell into three groups: masters colleges and universities with larger programs that awarded at least two hundred masters degrees; masters colleges and universities with medium programs that awarded one hundred to one hundred ninety-nine masters degrees; and masters colleges and universities with smaller programs that awarded fifty to ninety-nine masters degrees.

The classification scheme for baccalaureate colleges was based on the level of associate and bachelor's degrees awarded, the number of bachelor's degree majors produced in arts and sciences and other professional areas, and the extent to which graduate degrees were awarded in the same fields as undergraduate degrees.

The associates colleges category consisted of institutions whose highest degree was the associate degree. It also included institutions at which bachelor's degrees accounted for fewer than ten percent of all undergraduate degree offerings.

Special focus institutions were, with some exceptions, categorized as institutions offering at least eighty percent of undergraduate and graduate degrees in a specialized area. They were divided into thirteen subgroups based on offerings in a single field or sets of related fields.

Tribal colleges were considered to be minority-serving institutions that were controlled and operated by Native American tribes. They comprised thirty-two fully accredited colleges and universities authorized by Congress to serve as land-grant colleges. They were located either on or near to Native American reservations and served in excess of thirty thousand students!

While the Carnegie Classification System was not designed to rank higher education institutions, when college administrators began their annual strategic planning process, they routinely used the agency's reports to make comparisons between their institution and those considered to be actual or aspirational peers. Students, corporate recruiters and funding agencies also refer to Carnegie Classification System reports when making decisions about which college to attend or the best institutional program to support.

In the Carnegie Classification update published in 2019, postsecondary institutions that awarded bachelors, masters and doctoral degrees were grouped into two broad categories—a small number of institutions with large enrollments, and a large number of institutions with small enrollments.

That report was published by the Center for Postsecondary Research located at the Indiana University School of Education. The decision to transfer responsibility for publishing the Carnegie Classification Reports to Indiana University was made six years earlier by the Carnegie Foundation.

Because America's education system is decentralized, the quality of the postsecondary experience often varies from state-to-state and even within states. In order to ensure that institutions of higher learning remain fully compliant with guidelines established by Congress, all colleges and universities receiving federal funds for student aid are required to be fully-accredited. While the DoED is not directly involved in the accreditation process, it authorizes private agencies to work on its behalf in conducting periodic assessments of colleges and universities.

Any institution targeted for review is initially contacted by representatives from one of the private accrediting agencies and invited to work collaboratively with members of its review team to identify, discuss and agree on criteria and measurement standards to be used in the pending assessment process.

After that preliminary step is completed, the accrediting agency instructs the institution to undertake a self-study, paying particular attention to priorities assigned as targets for review and agreed upon in the initial consultation. Requirements for self-study generally include asking the college or university to measure performance in each assessment category; monitor institutional effectiveness in all performance areas; and identify programs that made significant and substantial improvements during the assessment period. After all relevant information has been collected, the institution is then instructed to submit a formal report to the accrediting agency that is fully supportive of its request for renewal of accreditation.

When the self-study report has been thoroughly reviewed, the accrediting agency contacts the institution and negotiates a mutually-convenient time for members of its team to visit the campus to conduct an on-site evaluation of programs and processes. On that occasion, members of the site-visit team meet with their counterparts from the institution's accreditation committee to listen to presentations from faculty and staff; review documents; inspect facilities; ask questions about programs and projects; and gather any information deemed useful for clarifying matters not adequately addressed in the self-study report. This process is usually guided by an itemized agenda that provides a realistic timeline for the site-visit team members to complete all of the assessment tasks.

Once concluded, team members prepare a comprehensive report that enables the agency to arrive at a final decision on the institution's request for renewal of accreditation.

If there is agreement that the institution's performance conforms to standards and practices authorized by Congress and established by DoED, then a decision is made to grant accreditation. If on the other hand, the reviewers determine that the institution has failed to meet expectations, then the request for accreditation is denied. The latter outcome can have a profoundly negative impact on a university's ability to operate, particularly with regard to accessing federal funding.

Few students are aware of the pivotal role that the Carnegie Classification System and accrediting agencies play in supporting and promoting postsecondary education in America. These data-driven organizations provide stakeholders with convenient access to detailed information on America's colleges and universities. Both agencies are to be commended for the invaluable contributions they make to protecting and preserving the quality of one of the world's most diverse assortment of higher education institutions.

CHAPTER 9

Funding Colleges and Universities

As indicated earlier, responsibility for funding the operation of America's colleges and universities rests with the states, not the federal government. Recognizing that access to high quality education is critical to workforce and economic development, state legislatures, together with local school districts, have enacted budget-planning processes that are responsive to the needs of their higher education institutions.

Annually, budget requests are developed by higher education administrators, submitted to the respective governor's office and forwarded to the state legislature for debate and approval. Spending plans typically include support for faculty and staff, sponsored research, financial aid, physical plant maintenance and upgrade, auxiliary services, and other key operational needs.

Departments such as Defense (DOD), DOE, DoED, Interior (DOI), Health and Human Services (HHS), and Agriculture (USDA), provide federal support for colleges and universities. Executive agencies such as the National Science Foundation (NSF), the National Aeronautics and Space Administration

(NASA), the Environmental Protection Agency (EPA), and the National Institute of Environmental Health Sciences (NIEHS) also provide funds to support research conducted by the nation's colleges and universities.

Two divisions within HHS—the NIH and the Centers for Disease Control and Prevention (CDC)—award billions of dollars annually to higher education institutions pursuing research that advances the mission of HHS to acquire knowledge that helps prevent, detect, diagnose and treat diseases.

It is estimated that in the decade of the 1990s, NIH appropriated over eighty percent of its annual budget to support more than fifty thousand competitive grants. Those awards were made to approximately three hundred thousand principal investigators affiliated with about two thousand five hundred universities, medical schools, and research institutions!

A typical NIH request for proposal (RFP) begins with a clear and comprehensive statement of purpose. It identifies the area of focus to potential applicants, providing them with background information on project goals, timetables, and specific aims for prosecuting the study. Criteria that determine whether an institution or investigator is eligible to submit a proposal are also delineated in the RFP, along with any special requirements associated with receiving the award.

There are special assurances that candidate institutions are instructed to provide indicating a willingness to support the project, and the capacity to fully comply with all rules and requirements specified by the agency funding the project. Recently, RFPs issued by NIH have emphasized the need for applicant institutions to indicate that successful prosecution of the proposed research project is likely to result in positive healthcare outcomes. Through its Clinical and Translational Science Awards, NIH was able to dictate how investigative research should be conducted in the new millennium.

The CDC is the other operating arm of HHS. Since its inception in 1946, it is best known for its role in disease prevention and control. As a branch of the U.S. Public Health Service, it allocates a significant percentage of its annual budget to funding research that is well aligned with the agency's interest in preventing the spread of infectious diseases.

NSF is an independent federal agency created by Congress in 1950. RFPs issued by this organization are similar to those already described for NIH. Categories of applicants eligible to respond to RFPs from NIH, CDC and NSF include colleges and universities, non-academic institutions, foundations, other non-profit organizations, corporations, state and local governments, and even individuals unaffiliated with any of the above-mentioned groups.

Federal departments and agencies also use a contracting process to acquire goods and services. These contracts are legally binding agreements that obligate the successful applicant to provide deliverables to the issuing agency as specified in the terms and conditions of the contract.

Big Pharma employs similar processes to satisfy corporate

R&D needs. The difference in the way Big Pharma makes funding decisions lies in the procedure used to identify, initiate, and prosecute a research project. For example, corporate executives target areas for R&D support based on the likelihood that the project will be financially successful. Typically, proposals are subjected to in-depth analysis during what is called a Project Operating Planning (POP) process. A committee composed of intramural scientists, managers, and external consultants conducts the POP review. Projects approved by this committee and authorized for funding by senior management have a designated principal investigator, project team, and annual budget.

When discovery research is initiated, project teams are usually small, comprising a limited number of biologists and one or more synthetic organic chemists. As work progresses, the team expands to include scientists from other specialty areas. This is referred to as the research (R) phase of the pharmaceutical drug discovery process.

To expedite matters, the project team generally relies on cell-based assays to model the disease under study. After proof of principle is established, a rodent model might be used to gather additional preliminary safety and efficacy data.

At every step of the discovery process, team members dedicate time and energy to keeping the project on track, technically, legally, as well as intellectually. It is not unusual for the project leader to solicit support from outside consultants using a variety of proprietary and non-proprietary strategies. What's most impressive about the overall process is the timeliness of its execution, and the commitment the corporation makes to ensure its success.

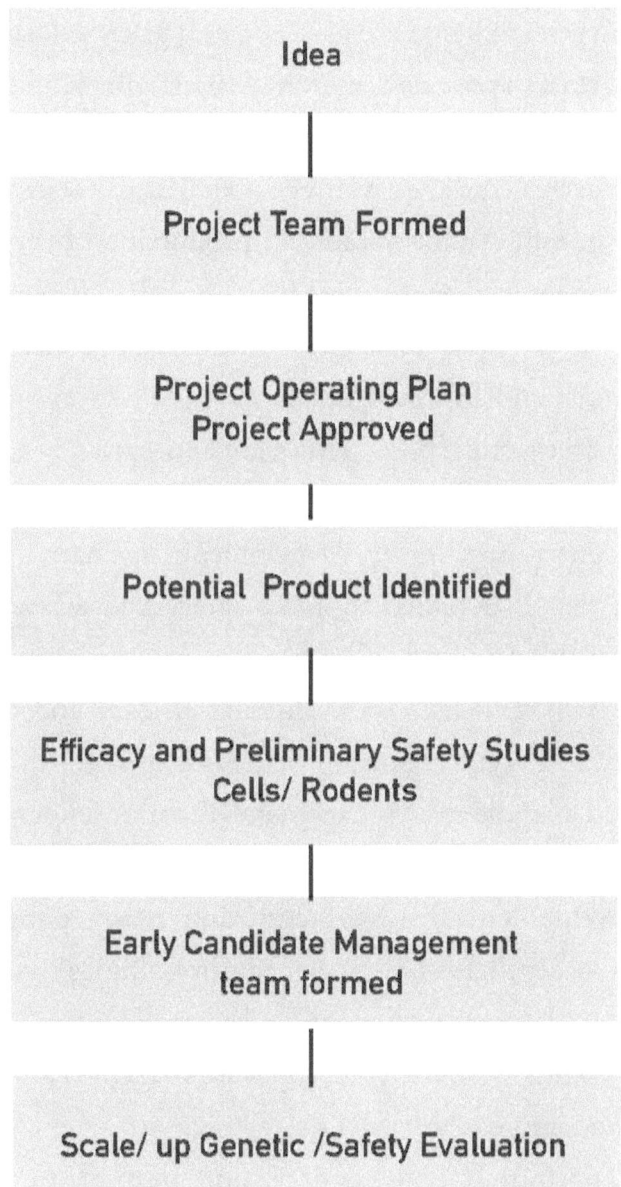

The 'R' of Pharmaceutical R&D

When all research data are gathered and analyzed, the project leader prepares a report documenting key findings and submits it to senior management for review and approval. If approval is granted, an Early Candidate Management Team is assembled and assigned responsibility for scaling-up production of the putative biological or chemical agent; preparing an IND (Investigational New Drug) application for FDA; guiding the project through the exploratory and full development stages; and ensuring that all relevant safety and efficacy issues are appropriately addressed, prior to initiating clinical development trials.

At the onset of clinical development, a Phase-I study is undertaken that is designed to assess safety and efficacy of the product in healthy subjects. If successful, then the process moves to Phase-II, which focuses on evaluating efficacy and dosing in patients. After safety and efficacy standards are met in Phase-II, an Advanced Candidate Management Team is empaneled and authorized to move the process into the full development (D) stage.

Full development involves recruiting many patients and launching a large (Phase-III) trial at multiple clinical sites. When this major trial is completed, all relevant data are collected, incorporated into a New Drug Application (NDA) that is filed for review and approval by the FDA.

When approval is granted, it is not unusual for everyone associated with guiding the project through the protracted R&D process to be invited to divisional or corporate headquarters to celebrate that accomplishment. These rare events enable biologists, chemists, toxicologists, pathologists, statisticians, clinicians, production managers and others to be recognized by their superiors.

```
Genetic/ Safety evaluation/
Drug Metabolism/ Quality
        Assurance
           |
Investigational New Drug (IND)
       Filed with FDA
           |
   Phase I Clinical Trials
    (Healthy subjects)
           |
Phase II Clinical Trials Dosing/ Efficacy
          (In patients)
           |
 Phase III Clinical Trials (In many
          patients)
           |
  New Drug Application (NDA)
        Filed with FDA
           |
    FDA Review/ Approval
     Product Launched
```

The 'D' of Pharmaceutical R&D

Full development culminates when the sales and marketing divisions schedule a date to officially launch the product for its primary indication. This event is usually held in a public setting replete with potential investors as well as members of the print and broadcast media.

The role that corporations play in funding America's colleges and universities is significant and substantial. Big Pharma is acutely aware that research universities are the primary sources of the innovative ideas industry relies on to replenish its robust pipeline of marketable products and processes. Additionally, Big Pharma knows that colleges and universities are the incubators from which they recruit their brightest and best employees. To satisfy these needs, corporate planning committees always try to ensure that there is adequate funding in their R&D budgets to support partnership initiatives with colleges and universities.

Grants are primarily used to solicit inputs from academics whose expertise is well aligned with corporate R&D interests. Many of these are awarded to universities using sole-source or competitive mechanisms designed to build relationships between industry and a select group of academic institutions.

The practice of offering consultancy agreements to academic scientists is one of the most pragmatic instruments Big Pharma relies on to develop relationships with highly-accomplished academic scientists. The specific language incorporated into the legal document drafted in support of such arrangements establishes a framework that allows Big Pharma to outsource projects that advance high-priority research agendas.

Similar arrangements are made when clinics, hospitals and large medical centers are contracted to conduct clinical trials on any novel pharmaceutical agent.

Another major resource that America's research universities and Big Pharma draw on to expand R&D budgets is the U.S. Patent and Trademark Office (USPTO). This agency, housed in the Department of Commerce, is the entity that institutions and research investigators turn to for help with protecting intellectual property.

The principal instrument that applicants seek from the USPTO is a patent. Patents are powerful legal documents authorizing inventors the sole right to exclude others from making, using or selling protected products or processes for a specified period of time.

U.S. colleges and universities also rely on endowments to grow their respective enterprises. Boards of Trustees establish and maintain endowments in a manner consistent with institutional guidelines and state law. Solicitations are coordinated through offices of institutional advancement or foundations. Major donors include alumni, corporations, foundations, institutes, wealthy individuals, employees and other interested parties.

A typical endowment is structured to protect the principal amount of the fund, while allowing investment income to be released as it accrues annually. Funds collected or pledged in response to well-organized capital campaign drives can be substantial. For example, with an estimated endowment of over forty billion dollars, Harvard is by far one of the wealthiest universities in the American higher education system.

A tally of funds awarded by government agencies and Big Pharma reveals that the U.S. allocates trillions of dollars annually in support of R&D activities. When grants awarded by foundations, wealthy individuals and other private sources are included, America's contribution to its higher education system is unmatched globally. This is a reality that everyone involved in science education should value and strive to preserve in perpetuity.

CHAPTER 10

Science for the Citizen

There is no single definition of science that captures the essence of what scientists do for a living. The Miriam Webster collegiate dictionary defines science as "the state of knowing: knowledge as distinguished from ignorance or misunderstanding." Academics tend to think of science as the study of all aspects of the physical, chemical, and biological world. Research investigators employed by Big Pharma consider science to be studying, developing, and applying knowledge gained from inquiries directed at advancing the human condition. In general, all attempts to define science usually include words like knowledge, truth, laws, methods, and other familiar nouns.

Middle school science students make no conscious effort to find out which definition of science is correct; instead, much of their classroom time is spent trying to gain a big-picture perspective of its various disciplines. This is the stage at which they are introduced to the universe and its planets. It's a time when their curiosity is piqued by studies of the earth's crust, its atmosphere, forests, oceans, rivers, and continents.

My introduction to science began in middle school where I learned about People and Homes in many Lands. I remember our lessons were focused on Eskimos and Pigmies. I was also taught to memorize the names of planets, as well as the position each planet occupied within the solar system.

During that period, it was not unusual for my textbooks to depict scientists as elderly white men seated at laboratory benches in long white coats, surrounded by round-bottom flasks filled with colorful solutions. I never saw myself reflected in any of those images! That false depiction of scientists and what they do could have stifled my interest in biology or chemistry—thankfully, it didn't!

When I entered graduate school, my mentor urged me to read a book entitled *Two Cultures* that was written by C. P. Snow. In his book, Snow posited that intellectual life in Britain was characterized by a split between two cultures—science, and the arts and humanities. He claimed that the schism in the British education system was so insidious that it stifled communication between members of the two intellectual communities, causing large segments of the society to remain illiterate with regard to basic science. Snow was confident that the problem could be corrected if the centralized British education system agreed to broaden student exposure to science during the primary and secondary years of schooling.

The two-culture phenomenon is a major problem in twenty-first century America where decentralization has created an education system in which each state determines how science is taught at the primary, secondary and tertiary levels.

For example, some American middle schools use traditional didactic methods and audiovisuals to introduce students to science, while others, because of budget limitations, offer students a less robust curriculum. In either case, science education in America, especially at the public school level, usually fails to excite student interest in biology, chemistry or physics. In spite of this reality, many college freshmen elect to major in a science discipline because of an interest in pursuing careers in medicine or dentistry.

During the years I spent as a college professor, I noticed that the majority of first year students chose biology as a major and identified medicine as their preferred career path. However, at the end of the freshman year, low grade point averages in science often forced many of them to change majors, opting for disciplines they considered to be less challenging academically.

My faculty colleagues were well aware of this problem. I remember that they often blamed the students, claiming that they did not invest the time and energy required to earn passes in freshman biology and chemistry. The administration on the other hand, tended to think that when students failed, much of the blame rested with the faculty assigned to deliver course content and grade exam papers. Regardless of who was at fault, by the end of the freshman year, student attrition rate was so high that it wreaked havoc on the science education pipeline.

What I learned from having taught science to freshman students is that those who managed to successfully navigate first-year biology and chemistry did so because they were properly introduced to the fundamentals of science during their high school years. Invariably, their love for science was sparked by a particular teacher whose commitment to the three pillars of science education (teaching, learning and applying) made the classroom experience consistently exciting.

The high school classroom is the place where interest in science is usually triggered. An exciting teacher, a well-designed curriculum, an interesting project combined with access to a dedicated mentor, are proven ways of awakening a student's curiosity about science. Outmoded lesson plans and inexperienced teachers assigned to deliver outdated course material can have the opposite effect. Unfortunately, science departments in too many American high schools are failing to address these obvious inadequacies.

In order to compensate for the lack of rigor in the way ninth-, tenth- and eleventh-graders are introduced to science, some American high schools employ master teachers to ensure that gifted students obtain adequate exposure to college level science courses before graduating. While this is viewed as a step in the right direction, only a tiny percentage of high school students benefit from such programs. Those that are not eligible to take college level courses during high school tend to think that science is a discipline reserved exclusively for the brightest and best students.

The approach commonly used to teach science in American high schools is referred to as the Layer Cake method because, on average, students are only required to take two science courses to satisfy graduation requirements. The problem with this approach to teaching science is that it limits student exposure to the fundamentals of basic science during the most critical period of their academic life.

A different approach to teaching science, and one that is generally practiced by nations that have centralized educational systems is referred to as the Spiral Staircase method. It requires students to take basic science courses during each of the four years they spend in high school.

An obvious advantage of using the Spiral Staircase method is that it exposes students to the fundamentals of basic science in their freshman year, and builds on that knowledge incrementally during the remainder of time they spend in high school. This stepwise approach to teaching science to high school students is very effective, and might explain why international students consistently outperform their American counterparts in tests designed to assess proficiency in STEM disciplines.

Laws guide much of the practice of science. These are usually based on observations believed to be universal and invariable. High school students often find the laws of biology, chemistry and physics difficult to comprehend. Many consider them archaic and dislike having to memorize them.

Theories generally represent beliefs that scientists develop about natural phenomena. These are usually data-driven and provide a basis for project planning and setting research priorities.

Hypotheses are different from theories in that they are based on tentative explanations of phenomena. Scientists generally draft hypotheses hoping that they can be proven experimentally. Then they formulate plans that enable them to generate the evidence needed to confirm or refute the claims embedded in their hypotheses.

I remember when I was a doctoral student, one of my thesis committee members told me I should invest as much as ninety percent of my time in crafting my hypothesis, and then devote the remaining ten percent to testing it! Although I considered that advice to be extreme at the time, I followed it rigorously and was often amazed by the success I had with executing an experimental plan that was well thought-out.

Irrespective of discipline, the narrative embedded in a well-crafted hypothesis shapes high quality discovery research. Specific aims should be identified delineating how each experiment will be conducted. Contingencies should be spelled out in the event that there is any need to modify the workscope of the project. Data collection, storage, and retrieval, as well as methods of analysis should be included in the planning process as a matter of common practice.

A well-designed research project usually begins with an investigator asking a simple question. During the implementation phase, as the project increases in complexity, multiple additional questions might be asked. This expands the timeline for completing the project. What is interesting is that even after engaging in a complicated and multidimensional experimental phase, most projects are concluded when the investigator arrives at a simple answer to the question initially posed!

While there is no formula that dictates who will be attracted to a career in science, student ability and enthusiasm are critical for academic success. These are buttressed by motivation, mentoring, positive socioeconomic indicators, and sometimes even luck.

Having the ability to recognize and nurture these qualities in students from every racial and cultural group is the best way for America to maintain and strengthen its science pipeline throughout the twenty-first century. A pragmatic approach to meeting this challenge would be to have the DoED or a study group acting on its behalf launch an initiative that prioritizes collection and storage of information from higher education institutions at every stage of the science education pipeline. In-depth analysis of these data should provide valuable insights into the student dropout phenomenon and what can be done to fix that nagging problem. The same stakeholder group could also be tasked with addressing questions related to science curriculum reform, in particular, and the impact of globalization on science education, in general.

Prevailing trends of corporate consolidation, combined with increasing numbers of companies moving operations off-shore, have devastated key sectors of the American economy. Wave after wave of plant closures, beginning with electronics, textiles, steel, and furniture, and continuing with pharmaceuticals, medical devices, and automobiles, can be cited as evidence of the ongoing contraction of America's manufacturing base.

Preserving the science education pipeline is a critical component of workforce development. Strengthening the bonds between industry and academia is unquestionably a national imperative.

CHAPTER 11

Teaching Versus Research

One of the primary missions of the American higher education system is to educate successive generations of the nation's citizens. Institutions that do this well consider teaching and research to be key elements of their continuing success. Doctoral research universities classified in the very high research category provide the best example of how teaching strategies should be deployed to enhance outputs from a resourceful and dedicated research faculty.

While teaching is not a major component of Big Pharma's mission, it is a resource that industry continues to draw on for professional development purposes as well as to keep their scientists connected to the academic world.

Learning how to balance teaching with research presents challenges that all colleges and universities, particularly the nation's HBCUs, confront on a daily basis. Deciding how to organize teaching faculty, allocate valuable research space, and compensate research-active faculty are matters that continue to present HBCU administrators with their greatest challenge.

I first became aware of the disparity in the way majority and minority institutions compensate research-active faculty when I became a professor at an HBCU. The state system in which I served routinely published the salaries paid to tenured professors. After perusing that list, it was clearly evident to me that there were significant differences in the way HBCU faculty were compensated compared to their counterparts at majority institutions.

I also learned that principle investigators conducting hypothesis-driven research at majority institutions were offered release-time from teaching and granted permission to use money allocated by the funding agency to supplement their annual salary. These perquisites were not available to me or any of the tenured professors recruited to serve as research program directors.

Overall, during my tenure as an HBCU professor, I spent as much time conducting hypothesis-driven research and training students as I did teaching. This does not mean that I neglected my teaching obligations. What it indicates is that I saw teaching and research as equally important for student development, and I considered the rewards of research—publications, patents and high levels of extramural funding—to be the most pragmatic way to increase student participation in basic and applied research.

I championed the issue of faculty involvement in organized research, even if the basic science departments provided no incentives for such participation. When I questioned department chairpersons and deans about their reluctance to embrace policies widely practiced by their counterparts at majority institutions, they often responded by claiming that state rules prevented them from granting release-time to research-active faculty. I considered this to be totally unacceptable. First, because I thought that guidelines for faculty compensation should apply equally to majority and minority serving institutions. And, second, I believed that if majority institutions were permitted to use funds provided by their foundations to support research-related activities, then HBCUs should be extended similar privileges.

With regard to my numerous requests to HBCU administrators to include hypothesis-driven research in the undergraduate science curriculum, they reasoned that adding a laboratory course would wreak havoc on degree administration, increasing tuition costs as well as the number of credit hours required for graduation. They also reminded me that HBCUs were established as teaching institutions and it was the duty of faculty to deliver on that important mission. What worried me most was when they told me that any student who wanted to pursue a career in research could satisfy that need by seeking admission to a doctoral degree-granting majority institution.

Although there is no Carnegie Classification Scheme for Big Pharma, I know from personal experience that America's pharmaceutical companies place a high premium on fostering and supporting hypothesis-driven research at every level of their operation. In spite of this well-documented fact, some academics are unusually harsh in their critique of the quality of basic research programs conducted by Big Pharma.

With limited understanding of the corporate R&D model, these critics believe that Big Pharma is not committed to conducting high-quality research. Because of this mistaken assumption, a small but significant number of college professors often discourage students from seeking employment at Big Pharma.

That happened to me, so I am intimately familiar with the problem. Shortly after completing postdoctoral training, I was cautioned by one of my mentors about accepting a job offer from industry. He felt that I would be compromising the quality of my doctoral and postdoctoral training if I went to work for Big Pharma. He characterized the science I would be exposed to in industry as "bucket" chemistry, and asked me to reject the offer without delay.

Fortunately, I agreed to their request for an interview and was so impressed by the scientists I met there, and the diversity of their research portfolios, that I had no difficulty casting my lot with Big Pharma.

Later in my career, I understood why my mentor was not enthusiastic about me seeking a job in industry. He was the quintessential traditionalist! He saw the university as a place where intellectual freedom was valued, and he wanted me to remain within academia, conducting research, and teaching the next generation of scientists.

After I became a college professor, I realized how difficult it was for science faculty to balance teaching and research responsibilities. Successfully competing for research funding and tenure placed severe constraints on research-active faculty at America's colleges and universities. This was in addition to the time many of them spent juggling full-time teaching assignments and grading students' exams.

The research culture in Big Pharma is different. While creativity and innovation are valued, research investigators spend little time on fundraising, a practice that allows them to focus exclusively on addressing the demands of their research projects.

It was in Big Pharma where I became aware of the linkage between basic and applied research, and the emphasis that is consistently placed on pursuing projects that address important biological problems. That exposure helped me to understand the schism that exists between academic and industry scientists. For example, academics speculate on how the discovery process in industry is initiated, and about the way corporate research is conducted. Many assume that without the traditional peer review process, project selection and implementation are lacking in rigor.

They claim that Big Pharma's primary reason for conducting research is to enrich shareholders, and they believe that preoccupation with greed inevitably compromises the quality of Big Pharma's research. They erroneously conclude that because the academic model for supporting basic research is not bottom-line driven, it is more likely to promote creativity and innovation than initiatives pursued purely for financial gain.

An overarching truth is that science is a business. Whether conceived in the halls of academia or by the planning committees of Big Pharma, good science is hypothesis-driven, and implementation is guided by goals and timetables. Funds are allocated to support discovery research only if reviewers are convinced that the goals of the project are realistic, and there is a high probability of achieving a successful outcome.

Whether support is derived from federal or corporate sources, expectations are the same—research that conforms to the mission and vision of the funding agency is all that matters.

The false impression that academic scientists hold about corporate research has stifled interaction between these two groups for decades. Academia's preference for publishing research findings compared to Big Pharma's predisposition to patenting products and processes, highlight stark differences between the two research cultures.

In the case of academia, pressure to publish is intense. It influences faculty behavior across departments and disciplines. The clarion call to "publish or perish" is heard on every university campus, and the louder the call, the more likely it is for professors to refrain from sharing research ideas with colleagues. They tend to retreat into silos, avoiding formal or informal interactions, particularly during the initial stages of a project. This practice of viewing colleagues with suspicion is prevalent in academia, and stems from the pressure placed on them by the tenure and promotion process to have their names listed first among the authors on peer-reviewed publications.

In contrast, corporate scientists prefer to patent their inventions rather than publish. Disclosure of ideas in peer-reviewed journals could invalidate any claim for novelty that an inventor might make to the USPTO to be granted exclusive rights to commercialize a particular discovery.

Because multidisciplinary teams conduct corporate research, complex projects are executed with amazing speed and efficiency.

Team members eagerly embrace their respective roles, contributing diligently to all aspects of the project's workscope.

These dissimilarities in the way science is practiced in academia and industry may explain why partnerships between academia and industry are difficult to initiate.

Throughout my years in Big Pharma, it was not easy for me to share information or exchange research materials with university colleagues. When I wanted to initiate such a relationship, I would have to consult with the licensing and development department and have one of their corporate lawyers prepare a formal agreement spelling out the conditions under which I would be permitted to interact with a prospective academic partner.

That secrecy agreement placed severe constraints on the proposed interaction. It limited material exchanges, and established rules of confidentiality. The legal department at the academic institution was equally demanding in its effort to protect the intellectual property rights of my presumptive partner.

While secrecy agreements and first right of refusal requests are legal instruments routinely used to initiate partnerships between industry and academia, they should never be seen as barriers to scientific collaboration. Instead, university administrators should view strategic partnership initiatives as indispensable tools for promoting faculty and student development initiatives.

It is ironic that in spite of the extraordinary successes resulting from unraveling the Human Genome sequence, few Americans are choosing science as a career path. A key reason for this has been the failure to recognize that teaching, learning, and applying are essential components of a well-designed roadmap for attracting and retaining science trainees.

Scientific concepts addressed in the classroom should be routinely supplemented with laboratory sessions that offer students hands-on opportunities to acquire, store, and analyze data. Successful science programs should focus on students doing science rather than having science done for them.

America's higher education institutions have failed to keep pace with the changing world of science for several decades. In an attempt to fix this problem, corporate America invested time, resources, and energy trying to find more creative ways to teach science. They have sponsored academic year and summer research internships; collected and analyzed student enrollment and retention data; and contributed to partnership initiatives that were directed at bringing science curricula into better alignment with workforce development needs.

The best models for promoting collaborations between universities and industry are the research parks and biotechnology hubs. Many of these centers of innovation were developed in the U.S. over the last forty years, when gene-cloning technologies resulted in the creation of hundreds of new companies. Cities like Boston, San Francisco, San Diego, Raleigh/Durham and many others became hot spots for biotechnology start-ups.

When I was a professor in the University of North Carolina System, I served on the board of the North Carolina Biotechnology Center (NCBC) and through that involvement, I was able to appreciate how that particular biotechnology hub contributed to the advancement of the state's workforce and economic development agenda.

The traditional model for university involvement in economic development is the technology transfer model. That mechanism enabled academic scientists to discover novel products or processes, patent their findings, and commercialize the respective inventions.

Because of its proximity to major universities located in Raleigh, Durham and Chapel Hill, NCBC through its Business and Technology Development Program, assisted start-up companies with financing, networking, marketing strategies, site location, business planning, venture capital, strategic partnerships, and technology assessment. The Center's Business Program utilized an Economic Development Investment Fund that provided low-interest loans to early-stage companies. That served as a proven mechanism for stimulating new business development.

There is no doubt that research parks and biotechnology hubs will continue to serve as centers of innovation, permitting America to expand its leadership role in science and technology throughout the twenty-first century.

CHAPTER 12

Science in the New Millennium

It is now sixty-seven years since James Watson stunned the scientific world with his incredibly prescient report on the physical structure of DNA. No one could have predicted that the young, brash American would be the final authority on a matter of such significance, not even the late, great organic chemists Edwin Chargaff, Linus Pauling or Robert Corey.

In the 1960s, when Nobel prize laureate Robert Holley showed how a novel class of molecules called tRNAs could assist with the translation of genetic information embedded in DNA, I was convinced that biochemistry was the right career path for me, and I elected to work with Holley's former student Dr. Jack Goldstein.

As Jack's graduate student, I isolated a novel species of RNA from bacterial cells and was consumed by the thought of pending celebrity. Later after I joined Big Pharma, my toolbox grew in size and sophistication enabling me to clone a mammalian gene, develop a recombinant DNA process for a food ingredient, and design and implement molecular strategies for discovering anti-cancer drugs. Each of these accomplishments heightened my conviction that discovery biology was even more exciting and rewarding than I had initially envisioned.

Currently, the genetic basis of life is taught in schools and colleges, and words like genomics, bioinformatics, metabolomics, microbiome, and proteomics have been incorporated into the language of science. DNA sequencing is faster and more accurate than when I first struggled to learn the classic Sanger method.

Technology has improved in ways that were unimaginable when scientists first proposed to sequence the human genome. That effort lasted ten years and cost three billion dollars. Today, a single human genome can be sequenced in twenty-four hours at a cost of approximately one thousand dollars!

Genetic engineering is the method of choice for livestock production, and genetically modified organisms (GMOs) are now commonly used to generate the food consumed by a majority of Americans.

Advances in genomics are also revolutionizing the practice of medicine, enabling clinicians to use a patient's genetic makeup to optimize drug discovery strategies (pharmacogenomics). One of the more recent developments is a new gene editing technology—CRISPR (clustered regular interspaced short palindromic repeats). This is offering seemingly unlimited possibilities to identify and correct abnormalities in a patient's DNA sequence. Analyzing DNA is now so commonplace that the general public can use DNA sequencing services to resolve paternity disputes or even trace ancestry.

As the third decade of the twenty-first century begins, and

in view of the promise that genomic technology offers to global economic development, America needs to take whatever action it deems necessary to protect one of its most precious resources—the doctoral research university. Just as Congressman Morrill fought to launch the land-grant colleges in the nineteenth century, bold congressional action is again needed to ensure that doctoral research universities are adequately resourced and well-positioned to maintain America's global leadership in science and technology.

Congress could ask its Committee on Education and Labor to undertake a comprehensive study of the nation's science education system. Data gathered from such a study could help decision makers craft a set of national standards for science curriculum reform; fix problems with teacher training and licensure; identify best practices for promoting and supporting partnerships between academia and industry; and target a select group of the nation's HBCUs for elevation to the Carnegie Classification level of doctoral research universities with very high activity.

Science curriculum reform is the first issue the Committee should address. Areas to be examined include course structure and content, lesson planning, textbook selection, and laboratory manual design. While upgrading the science curriculum is important, delivering the new lesson plans will only be effective if there are concurrent improvements in teacher preparation.

Teacher training and licensure is the second problem the Committee should make a concerted effort to tackle. As indicated earlier, the public school system is a state-run enterprise. Teacher training and certification are therefore state responsibilities. Changing the way public school teachers are trained and licensed is critical to the delivery of the modified science curriculum. A practical way to proceed would be to have the Committee recommend that the DoED work collaboratively with Schools of Education to upgrade criteria currently in use for degree administration and teacher certification.

In addition to taking courses, prospective science educators should be required to complete at least one internship before graduating. This internship should be conducted in a research-intensive setting, and under the guidance of an experienced investigator.

The new modality for teacher training combined with the introduction of the Spiral Staircase approach to delivering science course content will minimize challenges that twelfth graders face as they transition from public school to freshman year of college.

The third assignment the Committee should take on involves developing a framework that's best suited for promoting and supporting partnerships between academia and industry. As indicated earlier, industry scientists cultivate relationships with their counterparts in academia in order to gain access to unpublished data, new technologies, and sometimes even adjunct appointments in one of the university's basic science departments. College professors benefit from these exchanges in very specific ways. They gain access to some of America's best-resourced research facilities, and may even be extended consultancy agreements as well as faculty development grants. Creating formal mechanisms that promote and support such interactions will accelerate the pace at which major R&D programs are prosecuted.

Fourthly, the Committee should try to follow the example set by Congressman Morrill. This would involve asking Congress to create a special fund that would be earmarked to elevate a select group of the nation's HBCUs to the Carnegie Classification level of doctoral research university-very high research activity.

Based on the most recent Carnegie Classification System report, there are approximately one hundred thirty-one higher education institutions in America that are classified as doctoral research universities-very high research activity, and another one hundred thirty-five that are assigned to the category of doctoral research universities-high research activity. One very disturbing fact about these statistics is that no HBCU is listed among the Tier-I group of colleges and universities, and all of the top performing HBCUs, including those with well-established doctoral and professional schools, appear in the Tier-II category.

This is a problem that requires urgent national attention, and the Committee should make every effort to draw on the expertise of state and local education agencies, as well as HBCU administrators, to identify current institutional deficits and develop workable solutions for upgrading physical plants; enhancing the number and quality of doctoral degree offerings; expanding graduation rates, and strengthening all aspects of day-to-day operations.

Institutions that could be elevated to the Carnegie Classification level of doctoral research university-very high research activity within the first five years of such a program include FAMU,

Hampton, Howard, Jackson State, NC A&T, and Texas Southern. These are uniquely positioned to benefit from the proposed higher education reform initiative. As indicated earlier, some have land-grant status, and others are recognized for having exemplary records training minority students in STEM disciplines.

The Committee should also ask Congress to consider engaging in a second round of institutional upgrades. In this cycle, two HBCUs with well-recognized medical schools and companion doctoral degree programs (Morehouse School of Medicine and Meharry Medical School) and one additional institution noted for its outstanding contributions to the Veterinary Medicine community (Tuskegee University), should be considered for elevation to Tier-I.

Establishing multidisciplinary Centers of Excellence on the HBCU campuses identified above would add an important resource to the translational research initiatives already supported by NIH. These HBCU Centers would be funded to conduct research on minority health disparities, and serve as repositories for storing and disseminating information on diseases that disproportionately impact minority groups.

Elements critical to the success of these HBCU Centers of Excellence include creating reliable governance structures that strip away burdensome administrative red tape; partnering with Big Pharma on disease etiology and treatment; and establishing robust procedures for institutional advancement including patent acquisition, and grant and contract management.

The final issue the Committee should address before completing its task is the rising cost of college education. This is a national problem that has long-term ramifications for student recruitment and retention.

For decades, increasing tuition rates have adversely impacted enrollment in the nation's colleges and universities. To counteract this trend, some research universities have resorted to using digital technology as an option for providing an increasing number of students with greater access to undergraduate courses. In fact, the launch of massive open online courses by Harvard and MIT—two large, prestigious research universities—is an excellent example of how the landscape of American higher education has changed in the new millennium.

Obvious benefits include providing a significantly larger number of students with a low-cost college education; and enabling them to have the freedom to decide how, when, or where to access instructional materials. Because these developments will have important implications for the manner in which science education is delivered in the future, swift action by Congress is the best way to ensure that America's research universities remain true to their mission of teaching, research and service.

www.ingramcontent.com/pod-product-compliance
Lightning Source LLC
Chambersburg PA
CBHW020443220526
45464CB00002B/831